NEW GCSE SCIENCE

Separate Sciences
For Specification Units B3, C3 and P3

AQA

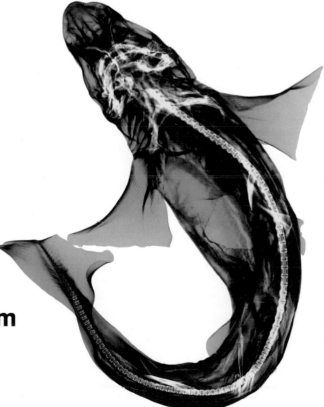

Series Editor: Ken Gadd

**Authors: Mary Jones,
Lyn Nicholls, Louise Petheram**

Student Book

William Collins' dream of knowledge for all began with the publication of his first book in 1819. A self-educated mill worker, he not only enriched millions of lives, but also founded a flourishing publishing house. Today, staying true to this spirit, Collins books are packed with inspiration, innovation and practical expertise. They place you at the centre of a world of possibility and give you exactly what you need to explore it.

Collins. Freedom to teach

Published by Collins
An imprint of HarperCollins*Publishers*
77 – 85 Fulham Palace Road
Hammersmith
London
W6 8JB

Browse the complete Collins catalogue at
www.collinseducation.com

©HarperCollins*Publishers* Limited 2011

10 9 8 7 6 5 4 3 2

ISBN-13 978 0 00 741459 8

British Library Cataloguing in Publication Data
A Catalogue record for this publication is available from the British Library

Commissioned by Letitia Luff
Project managed by Hanneke Remsing and 4science
Edited by 4science
Proofread by Life Lines Editorial Services
Indexed by Nigel d'Auvergne
Designed by Hart McLeod and Peter Simmonett
New illlustrations by 4science
Picture research by Caroline Green
Concept design by Anna Plucinska
Cover design by Julie Martin
Production by Kerry Howie
Contributing authors John Beeby, Nicky Thomas and Kiran Locke

Printed in China

Acknowledgements – see page 220

Contents

Biology

Chemistry

Physics

How to use this book

Welcome to Collins New GCSE Science for AQA!

The main content

Each two-page lesson has three levels:

> The first part outlines a basic scientific idea

> The second part builds on the basics and develops the concept

> The third part extends the concept or challenges you to apply it in a new way.

Information that is only relevant to the Higher tier is indicated with 'Higher tier'.

Each section contains a set of level-appropriate questions that allow you to check and apply your knowledge.

Look for:

> 'You will find out' boxes

> Internet search terms (at the bottom of every page)

> 'Did you know' and 'Remember' boxes

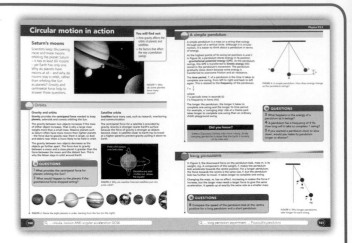

Units and sections

Each Unit is divided into two sections, allowing you to easily prepare for assessment.

Link the science you will learn with your existing scientific knowledge at the start of each section.

Checklists

Each section contains a checklist.

Summarise the key ideas that you have learned so far and see what you need to know to progress.

Exam-style questions

Every section contains practice exam-style questions for both Foundation and Higher tiers, labelled with the Assessment Objectives that they address.

Familiarise yourself with all the types of question that you might be asked.

Worked examples

Detailed worked examples with examiner comments show you how you can raise your grade. Here you will find tips on how to use accurate scientific vocabulary, avoid common exam errors, improve your Quality of Written Communication (QWC), and more.

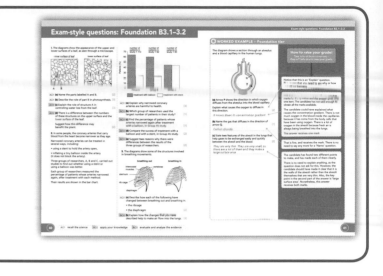

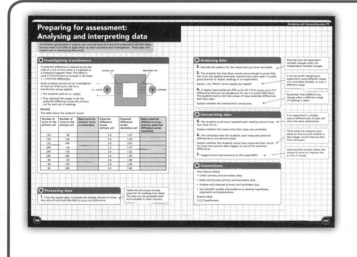

Preparing for assessment

Each Unit contains Preparing for assessment activities. These will help build the essential skills that you will need to succeed in your practical investigations and Controlled Assessment, and tackle the Assessment Objectives.

Each type of Preparing for assessment activity builds different skills.

> Applying your knowledge: Look at a familiar scientific concept in a new context.

> Planning an investigation: Plan an investigation using handy tips to guide you along the way.

> Analysing and interpreting data: Process data and draw conclusions from evidence. Use the hints to help you to achieve top marks.

Assessment skills

A dedicated section at the end of the book will guide you through your practical and written exams with advice on: the language used in exam papers; how best to approach a written exam; how to plan, carry out and evaluate an experiment; how to use maths to evaluate data, and much more.

Biology B3.1–3.2

What you should know

Dissolved substances

Diffusion is the spread of the particles of a gas, or of any substance in solution, that results in a net movement from a region where they are in a higher concentration to a region with a lower concentration.

Dissolved substances can move in and out of cells by diffusion.

Oxygen for respiration moves into cells by diffusion.

 Name one substance that moves out of a respiring cell by diffusion.

Exchange systems

Plants use carbon dioxide and produce oxygen during photosynthesis. The gases move in and out of the leaf through stomata.

Water vapour is lost through the stomata, during transpiration.

Aerobic respiration takes place continuously in plants and animals, so cells require a supply of oxygen.

In mammals, oxygen and carbon dioxide move in and out of the blood in the lungs.

Digested food substances move into the blood from the digestive system.

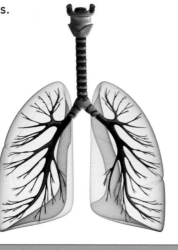 Name the process by which oxygen and carbon dioxide move in and out of the blood in human lungs.

Transport systems in plants and animals

Xylem and phloem are tissues that transport substances around a plant.

In mammals, the circulatory system transports materials, including oxygen, around the body.

Breathing rate and heart rate increase during exercise, to supply more oxygen to the muscle cells.

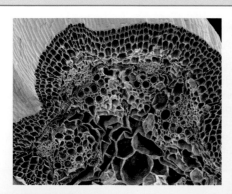 What type of blood vessel takes away blood from the human heart?

You will find out

Movement of molecules in and out of cells: Dissolved substances

> Dissolved substances move in and out of cells by diffusion or active transport.

> Water diffuses through partially permeable membranes by osmosis.

> Many organ systems are adapted to allow diffusion to take place rapidly.

Movement of molecules in and out of cells: Exchange systems

> Gaseous exchange in humans takes place across the alveoli in the lungs.

> Movements of the ribcage and diaphragm cause pressure changes that move air into and out of the lungs.

> Root hairs provide a large surface area for the uptake of water and mineral ions into plants.

> Carbon dioxide diffuses into plant leaves through stomata. Water vapour diffuses out.

Transport systems in plants and animals

>The human heart has four chambers. It pumps blood through arteries. Blood returns to the heart in veins.

> Blood is a tissue containing plasma, red cells, white cells and platelets.

> Red blood cells transport oxygen. White blood cells help to defend the body against microorganisms. Platelets help blood to clot.

> In plants, xylem tissue transports water and mineral ions from roots to leaves. Phloem tissue transports sugar from leaves or storage organs to all other parts of the plant.

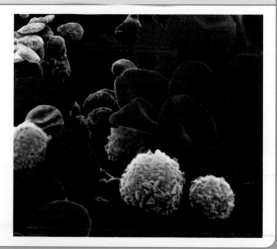

Osmosis

Kidney machine

When the kidneys stop working, blood can be cleaned in a kidney machine. The blood runs through tiny tubes made of a material like Visking tubing. It has microscopic holes that let some substances through but not others.

You will find out:

> how a partially permeable membrane works
> about osmosis
> Visking tubing is an artificial partially permeable membrane

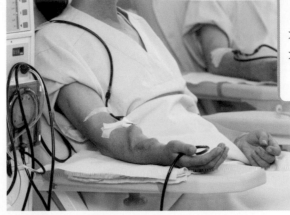

FIGURE 1: Kidney dialysis machines contain partially permeable membranes which filter harmful substances out of the blood.

Demonstrating osmosis

Visking tubing is a **partially permeable membrane**. This means it has holes in it that are big enough to let small molecules through, but not large molecules. Visking tubing will let water molecules through, but not sucrose molecules.

In the experiment in Figure 2, the Visking tubing contains a solution with a lot of sucrose molecules and also some water molecules. In the beaker, there is a solution with a lot of water molecules, and a few sucrose molecules.

The water molecules diffuse from the dilute sucrose solution into the concentrated sucrose solution. They do this because the 'concentration' of water molecules is greater outside the tubing than inside.

The level of liquid gradually moves up the glass tubing because of the water diffusing into the sucrose solution. The diffusion of the water molecules is called **osmosis**.

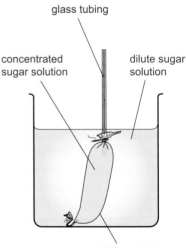

glass tubing

concentrated sugar solution

dilute sugar solution

Visking tubing – a partially permeable membrane

FIGURE 2: Using Visking tubing to demonstrate osmosis.

QUESTIONS

1 Copy and complete this sentence.

Osmosis is the diffusion of .... from a .... solution to a .... solution, through a .... membrane.

2 Explain what 'partially permeable' means.

How osmosis happens

Figure 3, on the next page, shows a concentrated sucrose solution separated from a dilute sucrose solution by a partially permeable membrane.

The sucrose molecules and the water molecules are all moving around. They bump into each other and then bounce off in different directions. Sometimes, they bump into the membrane. Sometimes, they meet a hole in the membrane.

If a sucrose molecule meets a hole, it just bounces back, because it is too big to get through. If a water molecule meets a hole, it passes through to the other side.

What will happen if the two solutions are left like this for a while?

> The sucrose all stays where it is.

> Some of the water molecules will move from one side of the membrane to the other.

There were more water molecules in the dilute sucrose solution than in the concentrated sucrose solution.

> More water molecules will move from the dilute solution into the concentrated one than in the other direction.

> Water will diffuse from the dilute solution into the concentrated solution.

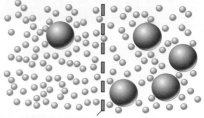

dilute solution — partially permeable membrane — concentrated solution

Key water molecule   sugar molecule

FIGURE 3: Explaining osmosis.

QUESTIONS

3 An experiment was carried out to find out how concentration affects osmosis. Several sets of apparatus as in Figure 2 were set up. They were all identical, except that the concentration of the sucrose solution in each one was different. The results are shown in the table.

Concentration of sucrose solution (g/dm³)	Height of liquid in the tube after five minutes (mm)
0	0
5	6
10	14
15	20
20	25

(i) Draw a graph to show these results.

(ii) Explain why the liquid went up the tube.

(iii) What do the results indicate about the effect of solute concentration on the rate of osmosis?

Isotonic drinks

When an athlete works hard they sweat, which cools their body. Sweat is mostly water, but it also contains ions such as K^+, Na^+ and Cl^-. A person doing exercise is also using up glucose, to provide energy for their muscle cells through respiration.

Some sports drinks are carefully balanced to contain the same concentration of water, ions and glucose as a person's blood should have. They are called **isotonic drinks**.

When the drink arrives in the person's small intestine, the water, ions and glucose quickly move by diffusion and osmosis into the blood, through the partially permeable membranes lining the gut. They help to restore the correct water and ion balance in the blood, as well as providing energy.

QUESTIONS

4 Most soft drinks contain water, sucrose and ions. What is special about isotonic drinks?

5 Suggest why drinking too many 'sports drinks' is not a good idea, unless you are using a lot of energy.

Q ... isotonic drinks

Osmosis and cells

You will find out:

> all cells are surrounded by a partially permeable cell membrane

> how water can move into and out of cells by osmosis

> how animal cells and plant cells are affected differently when they lose or gain water by osmosis

Storage solutions

Sometimes, a person needs a blood transfusion because they do not have enough red blood cells. Some of the blood that people donate is stored as 'red cell concentrate'. The red blood cells are separated out of the blood and then mixed with a solution containing salt and glucose. The solution has to be exactly the same concentration as the red blood cells, otherwise the red blood cells would be killed.

FIGURE 1: Blood packs in storage.

Osmosis and animal cells

Figure 2 shows an animal cell in distilled water. The cytoplasm in the cell is a fairly concentrated solution. The cell membrane is a partially permeable membrane.

Water goes into the cell by osmosis. The cell swells up with all the extra water. Eventually it bursts.

Did you know?

The National Blood Service says that 96% of us rely on the other 4% donating blood.

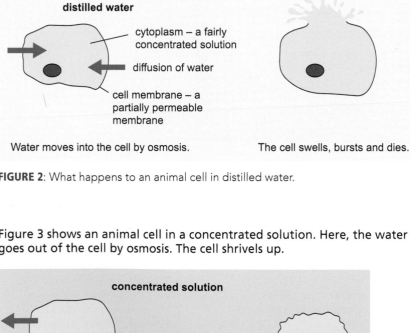

distilled water

cytoplasm – a fairly concentrated solution

diffusion of water

cell membrane – a partially permeable membrane

Water moves into the cell by osmosis.

The cell swells, bursts and dies.

FIGURE 2: What happens to an animal cell in distilled water.

Figure 3 shows an animal cell in a concentrated solution. Here, the water goes out of the cell by osmosis. The cell shrivels up.

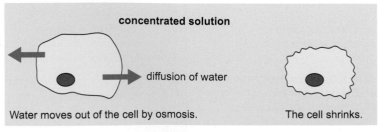

concentrated solution

diffusion of water

Water moves out of the cell by osmosis.

The cell shrinks.

FIGURE 3: What happens to an animal cell in a concentrated solution.

QUESTIONS

1 Copy and complete these sentences.

If an animal cell is put into pure water, the water goes .... the cell by .... This happens because the cell .... of the animal cell is a partially .... membrane.

2 Explain why animal cells shrivel up if you put them into a concentrated solution.

3 Red blood cells for transfusions are kept in a solution of salt and glucose, and not in pure water. Explain why this is important.

Osmosis and plant cells

Figure 4 shows a plant cell in distilled water.

Like animal cells, plant cells have a partially permeable membrane. They also have a cell wall, which is fully permeable.

Water goes into the cell by osmosis. The cell swells up. It does not burst, because it is surrounded by a strong cell wall.

Figure 5 shows a plant cell in a concentrated solution. Water goes out of the cell by osmosis. The inside of the cell shrivels up.

The cell wall is strong and does not collapse. The inside of the cell pulls away from the cell wall. Sometimes, the cell membrane tears when this happens and the cell dies. Sometimes, you can 'revive' the cell by putting it into water again.

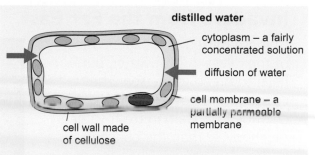

distilled water

cytoplasm – a fairly concentrated solution

diffusion of water

cell membrane – a partially permeable membrane

cell wall made of cellulose

Water moves into the cell by osmosis.

FIGURE 4: A plant cell in distilled water.

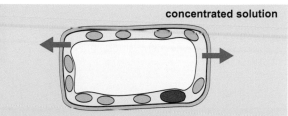

concentrated solution

Water moves out of the cell by osmosis.

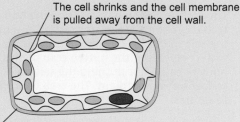

The cell shrinks and the cell membrane is pulled away from the cell wall.

The strong cell wall stays the same.

FIGURE 5: A plant cell in a concentrated solution.

QUESTIONS

4 Explain why animal cells burst if you put them into pure water, but plant cells do not.

5 Root hair cells are found near the tips of plant roots. They take up water from the soil. Explain how water moves into a root hair cell.

6 A student wrote, "In plant cells, water moves in and out through the cell wall by osmosis." What is wrong about this sentence?

Potato chips

A student investigated the effect of osmosis on potato cells. She cut three chips out of a raw potato. She measured the chips and recorded their lengths.

Then she put each chip into a different liquid. One went into distilled water, one into a dilute sucrose solution and one into a concentrated sucrose solution.

She left the chips for 30 minutes and then measured them again. Her results are given in Table 1.

TABLE 1: The effect of osmosis on potato cells.

Solution	Change in length (mm)
water	+ 6
dilute sugar solution	+ 1
concentrated sugar solution	– 4

QUESTIONS

7 Explain the results given in Table 1. (Think about what happens to the individual cells in the chips and how this will affect a whole group of them.)

8 Suggest how the student could alter her investigation (a) to make her results more repeatable and to check that they are reproducible (b) to find out the concentration of the cytoplasm in the potato cells.

Q ... osmosis AND plant cells

Active transport

Invaders from the Far East

Chinese mitten crabs spend most of their lives in freshwater, but have to breed in the sea. The concentration of sodium ions and chloride ions (salt) in freshwater is much lower than the concentration in the crab's body. When the crab is in freshwater, cells in its gills must pump these ions into its body from the water.

FIGURE 1: The Chinese mitten crab has invaded British rivers, accidentally carried here in ships' ballast tanks.

What is active transport?

The concentration of salt inside the body of a mitten crab is greater than in freshwater. You might expect the salt to diffuse out of the crab's body and into the water. However, the mitten crab pumps salt into its body because it is pushing the salt up its **concentration gradient**. The crab uses energy to do this.

> Active transport allows cells to absorb ions from very dilute solutions.

> Every living organism uses **active transport**.

For example, cells in plant roots use active transport to take up useful minerals, such as nitrate ions, from the soil – even when the concentration of nitrate ions in the soil is very low.

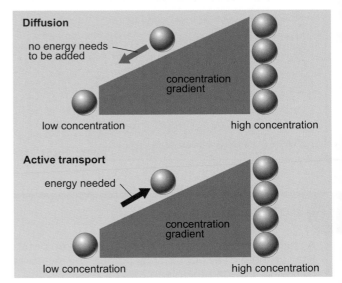

FIGURE 2: Active transport uses energy to move substances against their concentration gradient.

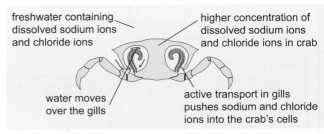

FIGURE 3: Crab gills use active transport to move dissolved ions into their cells.

QUESTIONS

1 State whether ions move up or down their concentration gradient in (a) diffusion (b) active transport.

2 The cells in the mitten crab's gills do not have to pump ions into themselves when the crab is in the sea. Suggest why.

Active transport and cell membranes

The part of the cell that does the active transport is the cell membrane. The membrane has special protein molecules in it that grab the ions. Then they use energy to move the ions across the membrane and into the cell.

The energy that the cell membrane uses to do this comes from respiration. Cells that do a lot of active transport have a lot of mitochondria. Inside the mitochondria, glucose is broken down and combined with oxygen to release energy.

QUESTIONS

3 Some carrot root cells were put into a solution containing three different ions, A, B and C. They were left in the solution for 30 minutes. The concentration of each ion in the solution, and inside the carrot root cells, was then measured. The table shows the results.

Ion	% concentration in the solution outside the root cells	% concentration inside the root cells
A	0.8	1.3
B	0.6	0.6
C	0.5	0.2

(i) Which ion has moved into the cell by diffusion?

(ii) Which ion has been moved out of the cell by active transport?

(ii) Which ion has been moved into the cell by active transport?

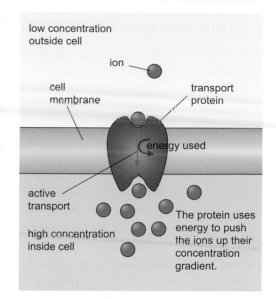

FIGURE 4: How cell membranes actively transport ions.

Did you know?

Fuelling active transport takes up to 40% of all the energy that a cell uses.

Root cells and active transport

Root hair cells, which absorb mineral ions from the soil, often use active transport. Their cell membranes can take in ions that the plant needs – such as nitrate ions or magnesium ions – from the water in the soil.

They obtain the energy for this by respiration. Roots therefore use a lot of oxygen, which they absorb from the air spaces in the soil.

QUESTIONS

4 Roots cannot photosynthesise. Where do they obtain the glucose that they need for respiration?

5 If it rains very heavily, soil may become waterlogged – so full of water that there are no air spaces in it. Suggest why many plants die if the soil in which they are growing stays waterlogged for several days.

FIGURE 5: Suggest why some of the trees have died.

Exchange surfaces

Gills for gas exchange

Unlike most amphibians, axolotls never come onto land. They live in lakes in Mexico. Their large, frilly gills help to absorb dissolved oxygen from the water.

FIGURE 1: Axolotls are like giant tadpoles that never grow up.

Exchange with the environment

Living organisms constantly:

> take in substances from their environment

> get rid of other substances.

Figure 2 shows some of the substances that humans exchange with their environment.

The surfaces that these substances must cross, to go into or out of the body, are called **exchange surfaces**. Two important exchange surfaces in the human body are:

> the lungs, where oxygen is taken in and carbon dioxide is given out

> the small intestine, where nutrients are absorbed into the blood.

QUESTIONS

1 What does the term 'exchange surface' mean?

2 Name two exchange surfaces in the human body.

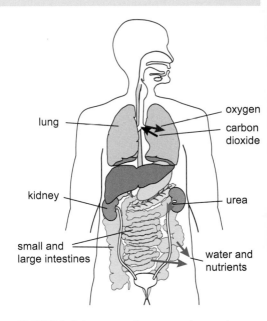

FIGURE 2: Substances that are exchanged between the human body and the environment.

The exchange surface in the small intestine

The inside of your small intestine is covered with thousands of tiny finger-like projections, called **villi**.

This is where all the nutrients from your digested food are absorbed into the blood. The villi help this to happen quickly because they have a huge surface area. If the intestine had a smooth lining, fewer nutrients would be able to move across at the same time – absorption would take much longer.

Remember

The plural of villus is villi.

Each villus contains a network of tiny blood capillaries. The blood is constantly on the move, carrying away the dissolved nutrients.

Some of the nutrients move into the villi by **diffusion**. They can do this if their concentration in the digested food, inside the intestine, is higher than the concentration in the blood.

The villi have very thin walls. This makes the diffusion pathway short.

Some nutrients, such as glucose, often have to be absorbed against their concentration gradient. The villi have to use active transport to make this happen.

To summarise, the lining of the small intestine is a good exchange surface.

> It has a **huge surface area**, produced by the thousands of tiny villi.

> The villi have **thin walls**, reducing the distance that substances have to diffuse.

> Villi contain **blood capillaries**, which carry away absorbed nutrients. This helps maintain the diffusion gradient between the inside of the intestine and the blood.

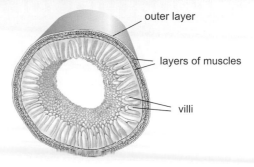

FIGURE 3: The small intestine has a huge surface area, made up of thousands of tiny villi.

Did you know?

The inner surface area of the small intestine is about 9 m² – the same as a king-size bed sheet.

QUESTIONS

3 What are villi, and where are they found?

4 Explain why it is important that villi have thin walls.

5 Suggest why the cells in villi contain a lot of mitochondria.

Absorption in a villus

Figures 4 and 5 show the internal structure of a villus.

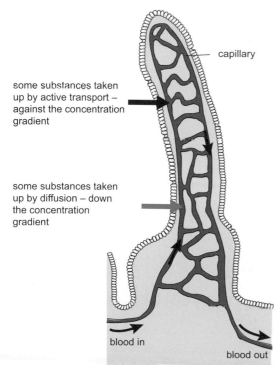

capillary

some substances taken up by active transport – against the concentration gradient

some substances taken up by diffusion – down the concentration gradient

blood in

blood out

FIGURE 4: How a villus absorbs nutrients.

FIGURE 5: This photograph of the lining of the small intestine was taken using a light microscope. A dark dye has been used to make the blood capillaries stand out.

QUESTIONS

6 Muscles, in the walls of the small intestine, keep the food inside it moving along. Blood, flowing in the capillaries inside the villi, keeps taking the absorbed nutrients away.

These two processes help to maintain a diffusion gradient for nutrients, so that they keep diffusing from the intestine into the blood. Explain how the processes maintain the gradient.

Preparing for assessment: Analysing and interpreting data

To achieve a good grade in science, you not only have to know and understand scientific ideas, but you need to be able to apply them to other situations and investigations. These tasks will support you in developing these skills.

✺ Storing sweet potatoes

Joel's mum bought sweet potatoes at the local market. Joel was looking forward to having them for dinner, but it turned out that he would have to wait. "I'm not using them yet," said his mum. "They're too fresh. I'll keep them for a few weeks so they get sweeter."

Hypothesis

Joel decided to investigate whether she was right. The hypothesis was that sweet potatoes would become sweeter a few weeks after harvesting.

He went back to the market stall at the end of the day, and asked if there were some leftover sweet potatoes that he could have. The stallholder gave him a bag of them, which he took to school.

Method

Joel took one of the sweet potatoes from the bag.

> He peeled it and then used a cork borer to cut out five cylinders from the potato. He cut each cylinder to exactly the same length.

> He dried each cylinder with paper towel, and then measured its mass.

> He placed one cylinder each in distilled water, or 5%, 10%, 15% or 20% sugar solution.

> After 20 minutes, Joel took the cylinders out of their solutions, dried them and measured their mass again.

> He calculated the change in mass of each sweet potato cylinder.

> Joel repeated his experiment after storing the sweet potatoes for 2 weeks, and again after 4 weeks.

Results

The table shows Joel's results.

Concentration of sugar solution (%)	Change in mass of potato cylinder (%)		
	at start	after 2 weeks	after 4 weeks
0	+ 2.4	+ 3.5	+ 4.3
5	− 0.2	+ 1.3	+ 2.5
10	− 1.8	− 0.4	− 0.4
15	− 2.6	− 0.3	− 1.1
20	− 2.9	− 2.2	− 1.4

Processing data

1. Joel decided to ignore one of his results, because he thought it was anomalous. Which result is anomalous?

> You could leave this question until after you have done question 3.

2. Suggest what Joel should have done about an anomalous result, rather than just ignoring it.

3. Draw line graphs to display Joel's results. Put *percentage concentration of sugar solution* on the x-axis, and *percentage change in mass of the potato cylinder* on the y-axis. The y-axis will need to have the origin (zero) somewhere in the middle, with increases in mass above this and decreases below it.

Draw a best fit curve for each of the three different storage times.

> Do not forget to deal with the anomalous result. Remember to label each of the three best fit curves that you draw.

Collecting data

4. Joel made several decisions about his method. Explain why each of these was a good idea:

(i) peeling the sweet potato before cutting out the cylinders

(ii) cutting the potato cylinders with a cork borer

(iii) drying the cylinders with a paper towel before measuring their mass

(iv) leaving the cylinders in the sugar solution for 20 minutes.

5. Suggest how Joel could have improved the investigation and checked the repeatability of his results.

> Think about variables. You could also think about repeatability – if Joel or someone else did the same investigation again, would they expect to obtain similar results?

> Suggest several ways, not just an easy one.

Analysing and interpreting data

6. Suggest the most important sources of error in Joel's investigation. For each one, suggest whether it would be random error or a systematic error.

7. Use your knowledge of osmosis to explain the shape of the curve that you have drawn for the potatoes at the start of the investigation.

8. Use Joel's results to decide whether his mum was right about sweet potatoes becoming sweeter after being stored for a while. Explain your answer.

> Suggest errors that really might have made a difference to Joel's results. Remember that sweet potatoes are living things, so there may be variation between them.

> You could start by explaining what made the cylinders gain mass, and what made them lose mass. Make sure that you use correct technical terms in your answer – including *osmosis*, *diffusion* and *partially permeable membrane*.

> Think about the shape of the curve. What can you deduce from the point at which the curve shows there would be no change in mass?

Connections

How Science Works

- Collect primary and secondary data

- Select and process primary and secondary data

- Analyse and interpret primary and secondary data

- Use scientific models and evidence to develop hypotheses, arguments and explanations

Science ideas

B3.1.1 Dissolved substances

Gas exchange

Suffocating water

Water can kill you, if it gets into your lungs. People drown in water because the water fills the air spaces in the lungs. This prevents oxygen from diffusing from the air into the blood.

FIGURE 1: How does water in the lungs prevent gas exchange?

The structure of the gas exchange system

In humans, gas exchange takes place in the **lungs**. The lungs are found in the upper part of the body, called the **thorax**.

The lungs are protected by the **ribcage**.

> When you breathe in, air moves down the trachea and into the lungs. Here there are millions of tiny air sacs called **alveoli**.

> Oxygen diffuses from the alveoli into the blood.

> Carbon dioxide diffuses in the other direction.

> When you breathe out, some of the carbon dioxide goes out of your body.

QUESTIONS

1 Put your finger on the nose in Figure 2. Trace the path that air takes on its way to an alveolus. List, in order, the structures that air passes through.

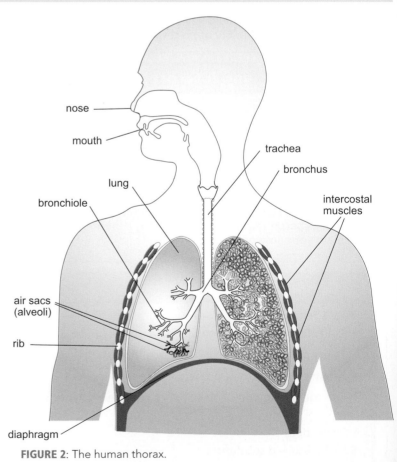

FIGURE 2: The human thorax.

Gas exchange

Gas exchange takes place across the surfaces of the alveoli.

Alveoli are very tiny, but there are thousands of them in each of your lungs. This provides an enormous surface area across which gas exchange can happen.

Each alveolus has blood capillaries pressed tightly against it.

> Oxygen diffuses from the air inside the alveolus and into the blood.

> Carbon dioxide diffuses in the opposite direction.

The constant movement of the blood, and the constant movement of air into and out of the lungs as you breathe, maintains a concentration gradient between the alveoli and the blood.

QUESTIONS

2 (i) List three features of the gas exchange surface in the lungs that help oxygen and carbon dioxide to diffuse across quickly.

(ii) For each of the features in your list, briefly explain their similarity with the exchange surface in the small intestine.

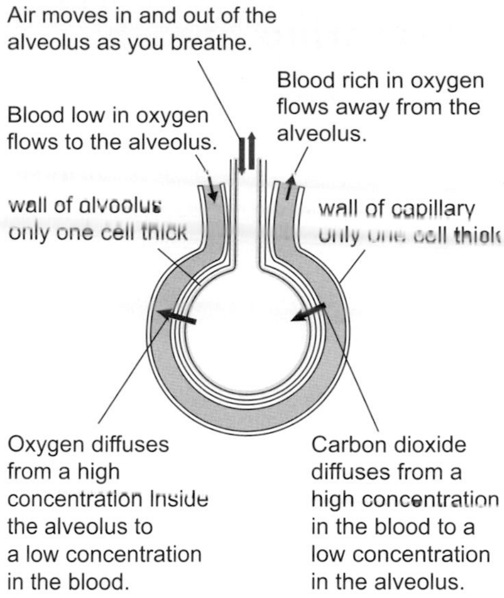

Air moves in and out of the alveolus as you breathe.

Blood low in oxygen flows to the alveolus.

Blood rich in oxygen flows away from the alveolus.

wall of alveolus only one cell thick

wall of capillary only one cell thick

Oxygen diffuses from a high concentration inside the alveolus to a low concentration in the blood.

Carbon dioxide diffuses from a high concentration in the blood to a low concentration in the alveolus.

FIGURE 3: Gas exchange between an alveolus and the blood.

Size and exchange surfaces

Some organisms are made of a single cell only. For these organisms, gas exchange is easy – they just allow oxygen to diffuse into them through their cell membrane, and carbon dioxide to diffuse out. They are so small that it takes almost no time for the gases to cross their surface and diffuse to or from every part of the cell.

> Small organisms have a large surface area compared with their volume.

> Humans have a large volume compared with their surface area.

The skin surface of large animals, such as humans, cannot exchange gases fast enough to supply all their cells. They have special gas exchange surfaces, made up of very thin surfaces, greatly folded to give a large area that speeds up exchange. There is also a transport system (such as the blood system) to carry oxygen to all the body cells and bring back waste carbon dioxide.

QUESTIONS

3 You are a terrestrial animal – you live on land, exposed to air that can be very dry. An axolotl is an aquatic animal.

(i) Explain why both you and an axolotl need special surfaces for gas exchange.

(ii) Suggest why an axolotl has gills on the outside of its body, but you have lungs tucked deep inside your body.

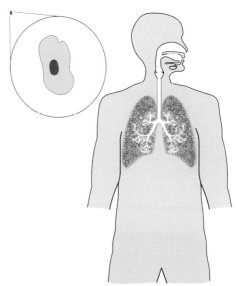

In a tiny organism, there is a lot of surface compared with its volume – it has a high surface area to volume ratio.

In a large organism, there is much less surface compared with its volume it has a low surface area to volume ratio. Folded exchange surfaces increase the surface area for exchange.

FIGURE 4: Large organisms need special gas exchange surfaces.

Breathing

Arresting breath

Police have to take care when they are restraining a person that they want to arrest. If someone is lying on their front and being pressed down, their thorax cannot expand as they breathe in. They might die unless correct procedures are followed.

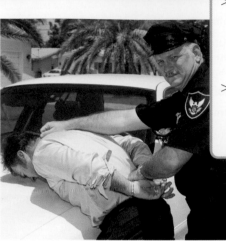

FIGURE 1: Police are trained in how to restrain violent people without restricting their breathing movements.

You will find out:

> muscles between the ribs and in the diaphragm move air into and out of the lungs

> muscle contraction reduces the pressure in the thorax, making air flow into the lungs

Inspired and expired air

Ventilation

The movement of air into and out of the lungs is called **ventilation**. Air is moved into and out of the lungs so that gas exchange can take place.

The air that you breathe in is sometimes called inspired air. The air that you breathe out is called expired air.

TABLE 1: The composition of inspired air and expired air.

Gas	% in inspired air	% in expired air
oxygen	21	16
carbon dioxide	0.04	4
nitrogen	78	78
water vapour	varies	saturated

Muscles for breathing

Lungs do not have muscles. Air is pulled into the lungs, and pushed out of them, by muscles that change the volume of the thorax.

There are two sets of muscles that help with this.

> One set is in between the ribs.

> The other set is in the diaphragm – the domed structure separating the thorax from the abdomen.

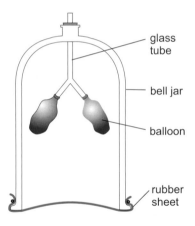

FIGURE 2: This model shows how the diaphragm helps in breathing. The rubber sheet represents the diaphragm. The balloons represent the lungs.

QUESTIONS

1 A student wrote, "Inspired air is oxygen, but expired air is carbon dioxide". Explain why this is incorrect.

2 Suggest why the percentage of nitrogen is the same in inspired air and expired air.

3 Predict what will happen if the rubber sheet, in Figure 2, is pulled down.

Did you know?

For every 10 years of life, from the age of 20 onwards, the total volume of air that you can move in and out of your lungs in one breath decreases by about 250 cm³.

How breathing works

The muscles involved in breathing work by changing the volume of the thorax.

When you breathe in:

> the muscles between the ribs contract, pulling the ribcage upwards and outwards

> the muscles in the diaphragm contract, flattening the diaphragm and pulling it downwards.

This makes the volume of the thorax larger. This reduces the pressure inside the thorax – and therefore inside the lungs. Air flows from the higher pressure area, outside the body, into the lower pressure area, inside the lungs.

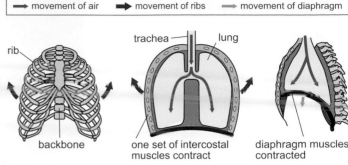

movement of air   movement of ribs   movement of diaphragm

rib   trachea   lung

backbone   one set of intercostal muscles contract   diaphragm muscles contracted

FIGURE 3: Breathing in (inspiration).

When you breathe out:

> the muscles between the ribs relax – the ribcage drops down again

> the muscles in the diaphragm relax – the diaphragm returns to its normal, domed shape.

This makes the volume of the thorax smaller, putting the air inside it under pressure. The air flows out from the high pressure area, inside the lungs, to the lower pressure, area outside the body.

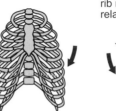
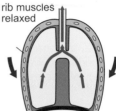

rib muscles relaxed

diaphragm muscles relaxed

FIGURE 4: Breathing out (expiration).

QUESTIONS

4 Describe the action of the diaphragm and the rib muscles when (a) breathing in (b) breathing out.

5 Explain why air moves into the lungs when the ribcage is raised.

6 How well does the model, in Figure 2, represent breathing in humans?

Breathing rate

The more oxygen that your body cells are using, the faster you need to breathe. Faster, deeper breathing provides a greater supply of fresh, oxygen-rich air to your lungs. It also takes away the carbon dioxide that is produced by the respiring cells.

You can control your own rate of breathing to some extent. Generally, though, it is controlled unconsciously.

> Signals pass along nerves from the brain to the muscles between the ribs and to the diaphragm muscles.

> These signals cause the muscles to contract and relax rhythmically, at just the right speed and with just the right force to supply sufficient oxygen to cells.

> The brain usually judges this by monitoring the concentration of carbon dioxide in the blood that flows through it.

QUESTIONS

7 Suggest why it is important for breathing rate to be controlled unconsciously.

8 Suggest some situations in which a person needs to control their breathing rate consciously.

Exchange in plants

Giant leaves

Gunnera manicata grows wild in Brazil. It has absolutely huge, prickly leaves. These leaves, like those of all plants, provide efficient exchange surfaces for the carbon dioxide that the plant uses in photosynthesis.

FIGURE 1: *Gunnera manicata*.

You will find out:

> carbon dioxide enters leaves by diffusion through stomata

> water and mineral salts are absorbed by roots

> root hairs increase the surface area of roots

> a flattened shape and air spaces increase the surface area of leaves

Plant exchange surfaces

During daylight, plants photosynthesise. They need to take in carbon dioxide and water. They produce oxygen, which they release to the air. Figure 2 shows how plants do this.

Leaves

Most plants have wide, flat leaves. This provides a large surface area: a lot of carbon dioxide and oxygen can diffuse in and out of the leaves very quickly.

Leaves are usually very thin. The diffusion distance from the air to the cells in the middle of the leaf is short.

Gases diffuse through tiny pores, called **stomata**, in the underside of the leaf.

Root hairs

Root hairs are found a little way behind the tips of each branch of the roots. Each root hair is part of a single cell, so they are very small. As there are thousands of them, they provide a large surface area for the absorption of water and mineral ions.

The water goes in by osmosis. The mineral ions go in by either diffusion or active transport.

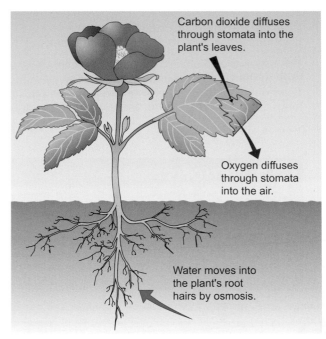

Carbon dioxide diffuses through stomata into the plant's leaves.

Oxygen diffuses through stomata into the air.

Water moves into the plant's root hairs by osmosis.

FIGURE 2: The exchange surfaces of plants.

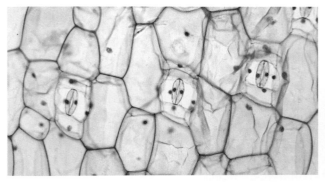

FIGURE 3: This is a magnified view of the underside of a leaf. There are tiny gaps (stomata) between the curved, red-stained cells.

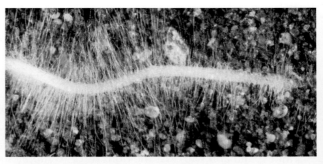

FIGURE 4: Root hairs provide a huge surface area for absorbing water from the soil.

Q ... plants AND minerals

Leaves and gas exchange

Figure 4 shows how carbon dioxide diffuses into the photosynthesising cells in the middle of a leaf.

The cells on the outside of the leaf, in the epidermis, fit closely together, forming a protective layer. Inside the leaf, the cells are not tightly packed. There are spaces in between them, which are filled with air. This means that many of the leaf cells are in direct contact with an air space, making it quick and easy for gases to diffuse into and out of the cells. These internal air spaces increase the surface area across which gas exchange can take place.

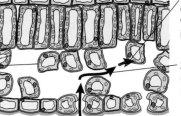

The surfaces of the cells next to the air spaces provide a large surface area for gas exchange.

3 Carbon dioxide diffuses across the cell walls and membranes, into the leaf cells.

2 Carbon dioxide diffuses through the air spaces inside the leaf.

1 Carbon dioxide diffuses through the stomata, down its concentration gradient.

FIGURE 5: How carbon dioxide diffuses into leaf cells.

Comparing exchange surfaces

You have now looked at four exchange surfaces – two in humans and two in plants. All of them are adapted for rapid exchange, and this means that they all have the same problems to solve in order to allow gases, or dissolved substances, to move into and out of the organism's body. Although these problems are solved in different ways in each instance, there are clear similarities between their structures.

Transpiration

Forests make clouds

In the cool air of the early morning, in a tropical jungle, the air is often full of clouds. The clouds form as water vapour evaporates from the leaves, cools and condenses into tiny water droplets. Eventually, some of these clouds will produce rainfall. If the trees are all cut down, there will be fewer clouds and less rain.

FIGURE 1: A steaming rainforest.

You will find out:

> plants lose water vapour from the leaves, mostly through their stomata

> evaporation from the leaves is faster when it is hot, dry and windy

> stomata can be closed by guard cells to reduce the loss of water vapour and prevent wilting

Transpiration

If you put a plastic bag over the top of a potted plant and leave it for a while, you will probably see droplets of water on the inside of the bag.

The water has come from the plant. Water vapour diffuses out of a plant's leaves, through its stomata. This is **transpiration**.

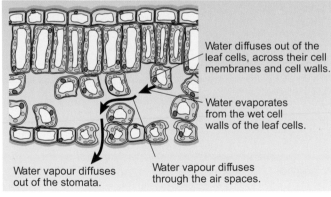

Water diffuses out of the leaf cells, across their cell membranes and cell walls.

Water evaporates from the wet cell walls of the leaf cells.

Water vapour diffuses out of the stomata.

Water vapour diffuses through the air spaces.

FIGURE 3: How water vapour is lost from a leaf.

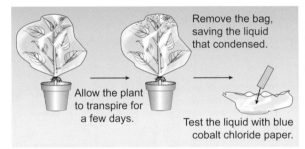

Remove the bag, saving the liquid that condensed.

Allow the plant to transpire for a few days.

Test the liquid with blue cobalt chloride paper.

FIGURE 2: Demonstrating the loss of water vapour from a plant's leaves.

QUESTIONS

1 What is transpiration?

2 Describe the pathway along which water vapour passes as it moves from the surface of a leaf cell and out into the air.

Benefits and drawbacks of transpiration

Controlling transpiration rate

Plants can partly control how fast transpiration happens by opening or closing their stomata. Each stoma is surrounded by two curved guard cells. These can change shape and close the hole between them.

FIGURE 4: How stomata can be opened and closed.

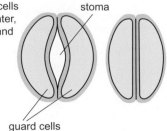

When the guard cells have plenty of water, they bend apart and open the stoma.

stoma

When the guard cells do not have much water, they become less curved and close the stoma.

guard cells

Benefits and drawbacks

> If it is hot, the evaporation of the water from the surfaces of the leaf cells has a cooling effect – like the evaporation of sweat from your skin.

> The loss of water from the leaves draws more water containing dissolved minerals up through the plant from the roots.

> If the plant loses too much water by transpiration, its cells may become short of water, causing the plant to wilt.

FIGURE 5: The effect of temperature, humidity and wind on the rate of transpiration.

Factors affecting transpiration rate

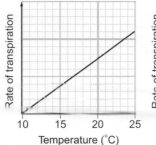
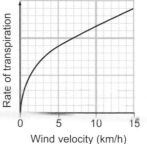
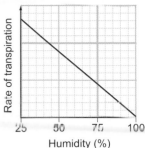

High temperature increases the rate of evaporation of water from the leaf cell surfaces. It also increases the speed of diffusion, because the water molecules have more kinetic energy. They move more quickly out of the leaf and into the air.

When it is windy, the wind takes away the moist air close to the leaf. This increases the diffusion gradient, and therefore increases the rate of diffusion.

When it is dry, there is less water vapour in the air. There is a steeper diffusion gradient between the moist air inside the leaf and the air outside it, increasing the rate of diffusion.

QUESTIONS

3 Three identical pot plants were placed on balances, as shown in the diagram.

The masses of the plants in their pots were measured over three days.

Day	Mass (g)		
	plant A	plant B	plant C
1	1325	1325	1325
2	1256	1289	1326
3	1184	1274	1325

Explain the results shown in the table.

Adaptations to different environments

Plant that live in different places have evolved different distributions and behaviour of stomata to allow them to survive.

For example, plants that live in hot, dry deserts generally open their stomata only at night. These plants tend to grow only very slowly.

FIGURE 6: Water lily leaves have more stomata in their upper surfaces.

QUESTIONS

4 In most plants, there are more stomata on the lower surface of the leaf than on the upper surface. Suggest how this helps the plant to conserve water.

Q ... desert adaptation stomata

The circulatory system

Heart transplant

This surgeon is holding a living human heart in his hands. The heart is going to be transplanted into someone who has incurable heart disease.

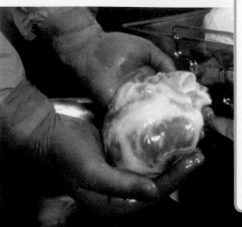

FIGURE 1: Why does a person die if their heart stops beating?

The human transport system

The human circulatory system transports substances from where they are taken into the body to the cells. The system also transports substances in the opposite direction.

> Oxygen is transported from exchange surfaces in the lungs to all the body cells. Body cells need oxygen for respiration.

> Carbon dioxide is produced by respiration. It is transported from the body cells to the lungs.

> Nutrients such as glucose and amino acids are transported from the small intestine to all the cells in the body.

These substances are transported in the blood in the circulatory system.

The circulatory system is made up of the heart and **blood vessels**. The heart pumps the blood, keeping it flowing through the vessels.

The heart

A heart is made almost entirely of muscle tissue. When it contracts, it becomes smaller. The muscle squeezes inwards on the blood inside the heart. This action pushes out the blood.

The heart has four chambers.

> The top two chambers receive blood. They are the **atria** (singular: **atrium**).

> The much larger chambers, at the bottom, pump blood out of the heart. These are the **ventricles**.

> Chambers on the left-hand side of the heart contain blood that has come from the lungs. This blood contains a lot of oxygen.

> Chambers on the right-hand side of the heart contain blood that has come from the body cells. This blood contains only a little oxygen.

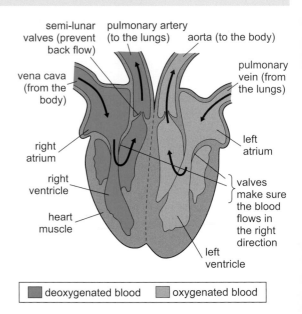

semi-lunar valves (prevent back flow) pulmonary artery (to the lungs) aorta (to the body)

pulmonary vein (from the lungs)

vena cava (from the body)

left atrium

right atrium

right ventricle

valves make sure the blood flows in the right direction

heart muscle

left ventricle

□ deoxygenated blood □ oxygenated blood

FIGURE 2: Inside the human heart. The heart is shown as though the person is facing you.

QUESTIONS

1 Name three substances that are transported by the circulatory system.

2 Name the four chambers of the heart.

3 Which chambers (a) receive blood from elsewhere in the body (b) pump blood out of the heart?

Q ... heart diagram

Heartbeat

Blood flows into the atria of the heart. From there, blood flows down through the open valves and into the ventricles.

The muscle in the walls of the ventricles then contracts. This increases the pressure on the blood.

> The blood is forced up against the valves between the atria and ventricles, forcing them shut. This stops the blood going back up into the atria.

> The blood is also forced against the valves in the big **arteries** – but these valves are made so that this pushes them open. Blood rushes out of the heart and into the arteries.

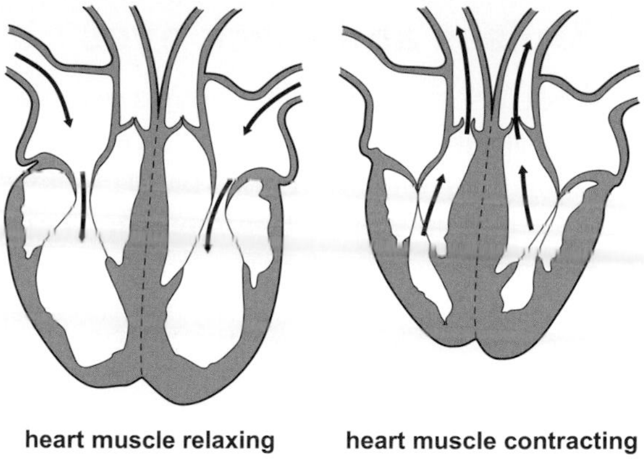

Two circulation systems

Blood leaving the right ventricle goes through the pulmonary arteries to the lungs. Here, it picks up oxygen before returning through the pulmonary **veins** to the left atrium.

Blood leaving the left ventricle goes through the **aorta** and then to all the other parts of the body. Here, it gives up a lot of its oxygen before returning through the veins to the right atrium.

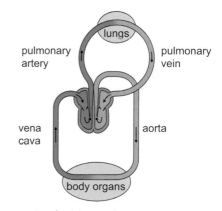

heart muscle relaxing   heart muscle contracting

FIGURE 3: How the heart pumps blood.

pulmonary artery   lungs   pulmonary vein

vena cava   aorta

body organs

FIGURE 4: The double circulatory system.

QUESTIONS

4 Explain how the valves in the heart stop blood flowing the wrong way, when the ventricles contract.

5 There are no gaps in the wall separating the right and left sides of the heart. How does blood travel from one side to the other?

Artificial hearts

There are not enough hearts available for transplants for everyone who needs one. Artificial hearts can keep people alive, for a short while. However, there are limitations with current artificial hearts.

> They have to be made from materials that are not attacked by the body's immune system and that do not decay.

> They use heavy batteries that are worn on the outside of the body.

It is much easier to make artificial heart valves. There are many people living normal lives today because their own damaged valves have been replaced with artificial ones.

QUESTIONS

6 Suggest why it is much easier to make successful artificial valves than a complete, working heart.

FIGURE 5: An artificial heart.

Q ... artificial heart ... pig heart valve transplant

Blood vessels

Bruises and black eyes

If you receive a hard knock on your skin, it can break some of the tiny blood vessels – capillaries – inside it. The blood leaks out and fills the spaces between the skin cells. At first it looks red, but then, as the blood is gradually broken down, it turns purple, then green and finally yellowish-brown.

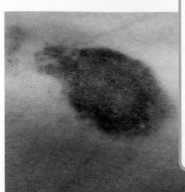

FIGURE 1: Bruises are the result of broken blood vessels.

Arteries, veins and capillaries

Blood flows around the body in tubes called **blood vessels**. There are three main kinds.

> **Arteries** carry blood away from the heart.

> **Capillaries** are tiny vessels that deliver materials close to every cell.

> **Veins** carry blood back to the heart.

Blood may leave the heart in the largest artery of all, the **aorta**. The aorta splits up into many tiny capillaries. These eventually join up to form a vein. The vein leads into the largest vein of all, the vena cava. This delivers blood back to the heart.

Some of the arteries that branch off from the aorta actually deliver blood to the muscle in the wall of the heart itself. These are the coronary arteries.

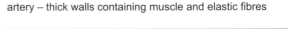

artery – thick walls containing muscle and elastic fibres

vein – thinner walls, containing some muscle and elastic fibres, also have valves

capillary – tiny, with walls only one cell thick

FIGURE 2: The three types of blood vessels.

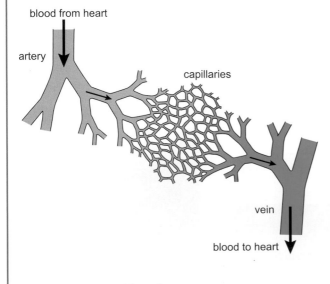

FIGURE 3: How blood flows from an artery to a vein.

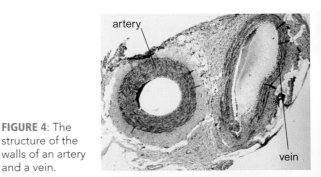

FIGURE 4: The structure of the walls of an artery and a vein.

QUESTIONS

1 Not all arteries carry oxygenated blood. Explain why.

2 Explain where the coronary arteries are, and what they do.

Structure and function

Arteries have to be strong, with elastic walls. They receive high pressure blood. It is also pulsing – each time the heart muscle contracts, a surge of blood rushes into the arteries. Artery walls are thick and strong with a lot of elastic tissue in them so that they can stretch and recoil as the pulses of blood move through.

By the time the blood reaches the veins, it is flowing smoothly and at a much lower pressure. Veins do not need thick, elastic walls. What they do need is valves to keep the blood moving in the right direction.

Capillaries are tiny. Their walls are only one cell thick. Many of them are just wide enough to allow a red blood cell to squeeze through. This means that they can deliver their contents really close to all the tissues in the body.

Blood can flow only one way through the valve – towards the heart.

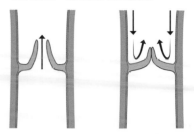

FIGURE 5: How the valves in veins work.

QUESTIONS

3 What exactly is causing the movement that you feel, when you take a pulse?

4 Copy and complete this table. Add several more features.

Feature	Arteries	Veins	Capillaries
direction of blood flow			
high or low blood pressure			
wall thickness			

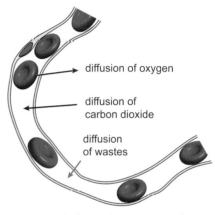

diffusion of oxygen

diffusion of carbon dioxide

diffusion of wastes

FIGURE 6: Exchange between capillaries and tissues.

Narrowing arteries

Sometimes, deposits of cholesterol, called **plaques**, build up in artery walls. This stiffens the walls, making it difficult for the artery to stretch as blood pulses through. It also reduces the space for blood to flow through.

This is particularly dangerous if it happens in the coronary arteries – the heart muscle will not get enough blood. The muscle cells cannot respire enough to provide energy to keep the heart beating. The person's heart becomes less effective. They may even have a heart attack.

One way of correcting this problem is to insert a little tube – a **stent** – into the artery. This keeps the artery open, allowing blood to flow through freely.

Figure 7 shows a stent in a coronary artery. The criss-cross tube in this X-ray is the stent. The X-ray does not show up the artery walls, only the blood inside – which is why the stent looks as though it is outside the blood vessel. It *is* inside the blood vessel.

QUESTIONS

5 An alternative way of treating coronary heart disease is to replace damaged coronary arteries with a vein taken from the leg. Suggest why using stents could be a better option for the patient.

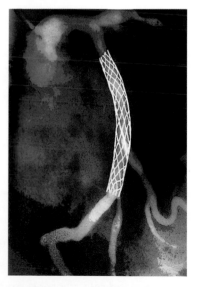

FIGURE 7: X-ray showing a stent inside a coronary artery.

Q ... heart disease GCSE

Blood

Blood with antifreeze

The Antarctic ice fish does not have any haemoglobin in its blood – it does not need it. At the very low temperatures where it lives, enough oxygen just dissolves in its blood plasma. It does have chemicals called glycoproteins in its blood. These keep the blood liquid even when the fish's body temperature is below 0 °C.

FIGURE 1: The Antarctic ice fish is adapted to live in icy cold water.

You will find out:

> blood is a tissue, containing red and white cells and platelets floating in liquid plasma
> how blood transports carbon dioxide, oxygen, nutrients and urea
> white blood cells fight microorganisms and platelets help blood to clot

Blood components

The liquid part of blood is the **plasma**. It is a pale yellow colour. Blood looks red because it contains huge numbers of **red blood cells**. Red blood cells contain a red chemical called **haemoglobin**.

Blood also contains **white blood cells**. These help to defend the body against microorganisms.

There are also tiny fragments of cells called **platelets**. These help blood to clot at the site of a wound.

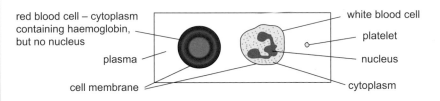

FIGURE 2: The components of blood.

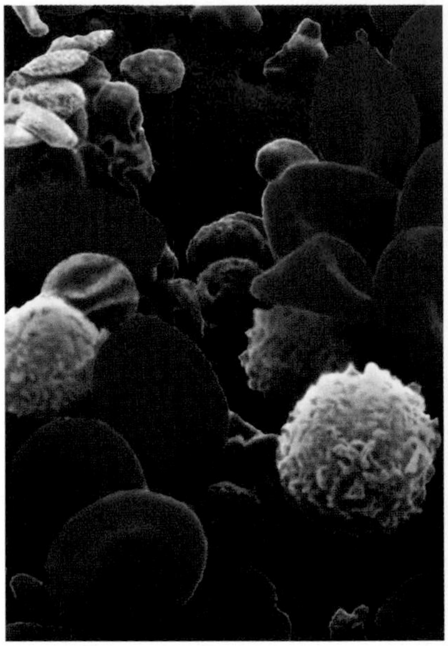

FIGURE 3: This photograph of blood was taken using an electron microscope. The white cells look like spiky balls. The small pink objects are platelets.

QUESTIONS

1 Which types of blood cells contain a nucleus?

2 Why does blood look red?

3 Look at Figure 3. Which type of blood cell is most abundant?

Transport by the blood

Blood transports many different substances around the body.

Transporting oxygen

Red blood cells transport oxygen. Their red pigment, haemoglobin, helps with this. When the blood passes through a capillary next to an **alveolus** in the lungs, oxygen diffuses into the red blood cells. The oxygen combines with the haemoglobin, forming a bright red substance called **oxyhaemoglobin**. The blood is now said to be **oxygenated**.

... blood vessels

When oxygenated blood flows through a capillary close to respiring cells, the oxygen leaves the oxyhaemoglobin and diffuses out of the blood and into the cells.

The oxyhaemoglobin splits up into haemoglobin and oxygen and the blood becomes less bright red. The blood is now said to be **deoxygenated**.

Transporting nutrients

Nutrients from digested food are absorbed into the blood in the small intestine. They include glucose, amino acids, mineral ions and vitamins. These dissolve in the blood plasma and are carried to all the other body organs.

Transporting carbon dioxide and urea

Cells produce waste substances. All cells produce carbon dioxide as a waste product of respiration. Liver cells produce **urea**, which they make from excess amino acids.

Carbon dioxide and urea dissolve in blood plasma, which carries them all over the body. When the blood passes through the lungs, the carbon dioxide diffuses out of the blood into the alveoli and is breathed out. When the blood passes through the kidneys, the urea is removed and is passed out of the body in urine.

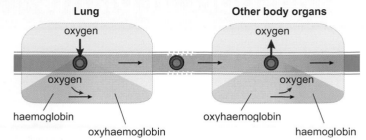

FIGURE 4: Transport of oxygen by the blood.

The cytoplasm contains a lot of haemoglobin, which combines with oxygen in the lungs and releases it when close to respiring cells.

There is no nucleus, making more room for haemoglobin.

The cell is very small, so it can fit through the very smallest capillaries.

FIGURE 5: How a red blood cell is adapted for transporting oxygen.

QUESTIONS

4 Copy and complete this table.

Substance	Where it is transported from	Where it is transported to	Part of the blood that transports it
oxygen			
soluble nutrients			
carbon dioxide			
urea			

Blood as a tissue

A tissue is a group of cells with similar structure and function.

Blood is usually classified as a tissue. It is an unusual tissue, firstly because it is liquid, and secondly because it contains different kinds of cells. Moreover, these cells perform different functions.

However, all of the different kinds of blood cells are made from the same kind of stem cells. These are found in the bone marrow.

QUESTIONS

5 The tissue lining the trachea and bronchi is another example of a tissue containing different types of cells. What are these cells, and what is their function?

6 Red blood cells live for only about 120 days, so thousands of millions of new ones are made each hour by the stem cells in the bone marrow. Suggest why red blood cells have a shorter life than most other cells.

Transport in plants

You will find out:

> about transport systems in plants

> how water moves from a plant's roots to its leaves

> how to use a potometer

Early plants

The first plants that lived on land were very small. Until they evolved tubes to carry water from the ground up through their bodies, plants could not grow tall. The first plant with water-carrying tubes was a club moss. It dates back about 400 million years. Club mosses can still be found growing today.

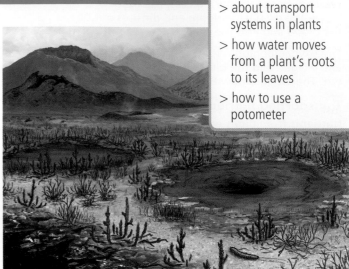

FIGURE 1: Earth may have looked like this, 400 million years ago.

Two transport systems

Plants have two transport systems. Both systems consist of long tubes, made of many cells joined end to end.

Xylem tissue transports water and mineral ions from the roots to the stem and the leaves.

Phloem tissue transports dissolved sugars from the leaves to all other parts of the plant.

QUESTIONS

1 Name the two tissues in a plant that help with transport.

2 Think back to what you have learned about how plants absorb and lose water.

(i) Which cells in the root absorb water?

(ii) How is water lost from a plant's leaves?

3 Explain why sugars need to be transported from leaves to roots.

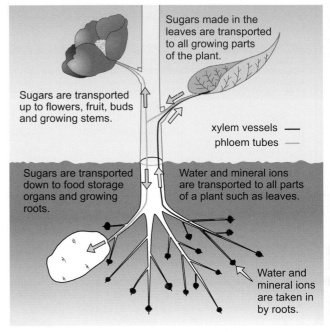

Sugars made in the leaves are transported to all growing parts of the plant.

Sugars are transported up to flowers, fruit, buds and growing stems.

xylem vessels ——
phloem tubes ——

Sugars are transported down to food storage organs and growing roots.

Water and mineral ions are transported to all parts of a plant such as leaves.

Water and mineral ions are taken in by roots.

FIGURE 2: The two transport systems in a plant.

The transpiration stream

Unlike mammals, plants do not have a heart to pump fluids through their transport systems. How do they keep things moving?

Water moves through xylem because of transpiration. Transpiration is the loss of water (as water vapour) from the leaves, as it evaporates from the cells and diffuses out into the air.

> The loss of water in the leaves reduces the pressure in the xylem vessels in the leaf.

> The pressure at the top of a xylem tube reduces.

> Water moves up through the tube, flowing from high pressure (at the bottom of the tube) to the top.

It is easy for the water to move through xylem, because xylem tubes are completely hollow. Xylem is made of the remains of dead cells – just their outer walls, with nothing at all inside them. The cells are stacked end to end, like long drainpipes.

The movement of water from the roots, through the xylem and out of the leaves is the **transpiration stream**.

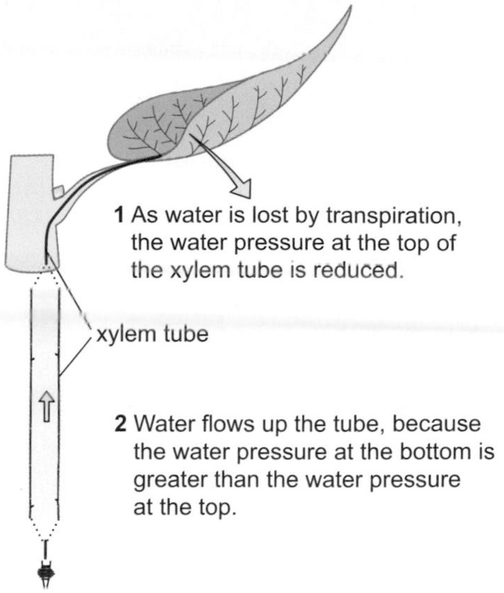

Using a potometer

You can use a piece of apparatus called a potometer to measure how quickly water is taken up and lost by the shoot of a plant.

To use the apparatus, you introduce an air bubble into the capillary tube. As water moves into the shoot and evaporates from its leaves, the air bubble moves towards the plant.

By measuring how fast the air bubble travels, you can get an idea of how fast the plant is transpiring. You refill the capillary tube (using the reservoir tap) to repeat the measurements.

If you put the apparatus where it is warmer, windier or less humid, you can find out how these factors affect the rate of transpiration.

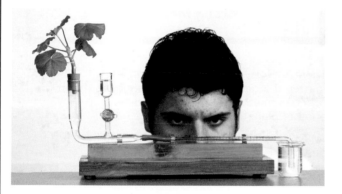

FIGURE 4: A potometer.

1 As water is lost by transpiration, the water pressure at the top of the xylem tube is reduced.

xylem tube

2 Water flows up the tube, because the water pressure at the bottom is greater than the water pressure at the top.

FIGURE 3: How water moves up xylem tubes.

QUESTIONS

4 Explain what makes water move up through the xylem.

5 Explain why water moves more quickly through the xylem on a hot day than on a cold day.

6 Most of the water that is taken up by a plant's roots is lost through transpiration. However, some of it is used up in a chemical reaction in the leaves. What is this reaction?

Moving across the root

Plants absorb water into their root hairs, by osmosis. Before the water can enter a xylem tube, it must travel from the outside of the root to the centre. Figure 5 shows how it does this.

QUESTIONS

7 Water moves into and across the root by osmosis. Explain why the movement of water in the xylem is *not* osmosis.

8 Explain how transpiration helps to keep water moving, into the root hairs.

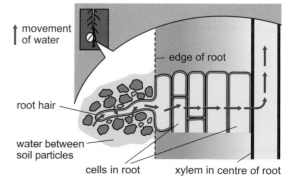

movement of water

edge of root

root hair

water between soil particles

cells in root

xylem in centre of root

FIGURE 5: How water is absorbed and transported in a plant

Preparing for assessment: Applying your knowledge

To achieve a good grade in science, you not only have to know and understand scientific ideas, but you need to be able to apply them to other situations and investigations. These tasks will support you in developing these skills.

✳ Giant machine

Andy is eleven years old. His class has a new science teacher and she has some different ideas for the lessons.

In one lesson, she gives each group of students a large sheet of paper and tells them that she wants them to design a machine. They can talk about their ideas and draw what they think will work. She tells them what the machine has to be able to do and then puts the list on the board for them to refer to. Look at the board in the picture to see the list.

> **Your machine must be able to:**
> - Get water from under the ground – though it can't move from one place to another.
> - Pump the water up high in the air. It doesn't need to store it there – it can just be released.
> - Do this using solar power.
> - Support itself and withstand fairly strong winds.

The class gets to work.

After they have finished, the students put their drawings up on the walls and look at each other's work. There are some weird ideas, but some of the features are quite similar.

"Now," says the teacher, "does anyone know what it is that you have actually designed?"

There are lots of puzzled looks and a few muttered suggestions. Someone suggests that it may be something to make life possible on another planet.

"Well," explains the teacher, "what you have just designed is a tree."

This is the drawing that Andy's group produced.

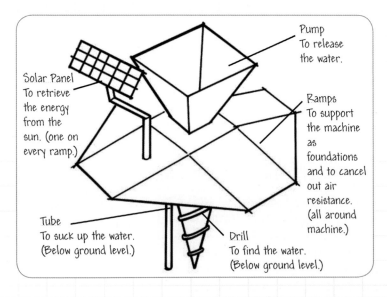

Solar Panel
To retrieve the energy from the sun. (one on every ramp.)

Pump
To release the water.

Ramps
To support the machine as foundations and to cancel out air resistance. (all around machine.)

Tube
To suck up the water. (Below ground level.)

Drill
To find the water. (Below ground level.)

✺ Task 1

(a) How does a tree obtain the water that it needs?

(b) Why does a tree need water?

✺ Task 2

(a) How does a tree get water to rise several metres (depending on the height of the tree)?

(b) Explain the energy transfers that take place in this process.

✺ Task 3

Look at the drawing that Andy's group produced. What are the similarities between their machine and a tree?

✺ Task 4

A tree also carries out other functions that the machine does not.

Describe three other functions.

✺ Task 5

The activity that Andy's teacher had set them was designed to make the class think about why trees are like they are.

Do you think it was a good way of doing this? Do you think that, by the end of the lesson, the group would have a better idea why trees have their various features?

✺ Maximise your grade

Answer includes showing that you...
recognise that plants need a supply of water.
recognise that the word equation for photosynthesis applies to a tree.
relate the uptake of water to the word equation for photosynthesis.
can describe how water moves through the tree.
can explain how energy is transferred to the water in transpiration.
recognise similarities and differences between the machine and a tree.
can use the machine to explain the form and function of various parts of the tree.
can draw together ideas to show how a design activity can give insights into the form and function of a tree.

E · C · A

Checklist B3.1–3.2

To achieve your forecast grade in the exam you will need to revise

Use this checklist to see what you can do now. Refer back to the relevant topics in this book if you are not sure. Look across the three columns to see how you can progress.

Remember that you will need to be able to use these ideas in various ways, such as:

> interpreting pictures, diagrams and graphs

> applying ideas to new situations

> explaining ethical implications

> suggesting some benefits and risks to society

> drawing conclusions from evidence you are given.

Look at pages 188–209 for more information about exams and how you will be assessed.

To aim for a grade E	To aim for a grade C	To aim for a grade A
Recall that water can move from cell to cell by a process called osmosis.	Understand that osmosis is the diffusion of water from a dilute to a more concentrated solution.	Explain how concentration differences affect the movement of water across a cell membrane.
Recall that, when exercising, people lose water and ions in sweat.	Describe how people lose water and ions in sweat.	Explain the benefits of isotonic drinks.
Recall that active transport requires energy from respiration.	Explain that active transport is movement against a concentration gradient.	Explain the active transport of ions across a cell membrane.
Recall two important exchange surfaces in humans.	Describe the features that increase the effectiveness of exchange surfaces.	Describe in detail how the surface area of the small intestine is increased to make absorption more effective.
Recall the key structures of the human gas exchange system.	Describe the process of gas exchange between the bloodstream and the air.	Explain (using the concept of surface area and volume) why large animals cannot exchange gases via their body surface.
Recall that the movement of air into and out the lungs is called ventilation.	Describe the movement of the ribcage and diaphragm in breathing in and out.	Explain inspiration and expiration with reference to the volume of the thorax and the pressure of air.

To aim for a grade E

Recall how plants absorb carbon dioxide and water.

Recall how plants lose water vapour from their leaves.

Recall the basic components and function of the circulatory system.

Describe the walls of arteries, veins and capillaries.

Recall each of the four main components of blood.

Recall the two transport systems found in plants.

To aim for a grade C

Explain how leaves and roots are adapted to their role as exchange surfaces.

Describe the benefits and problems of transpiration.

Describe the structure of the human heart.

Understand the importance of stents, particularly with reference to the coronary arteries.

Describe the function of each component of the blood.

Describe the role played by the xylem and phloem in transport of materials around plants.

To aim for a grade A

Explain how the different arrangement of cells within a leaf aids gas exchange.

Explain in detail how environmental factors affect the rate of transpiration.

Explain how the action of the atria, ventricles and valves maintains the flow of blood through the heart.

Explain the role of haemoglobin in the transport of oxygen by the blood.

Explain the movement of water from the soil into the xylem vessel.

1. The diagrams show the appearance of the upper and lower surfaces of a leaf, as seen through a microscope.

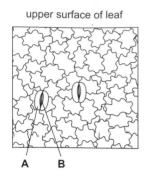

upper surface of leaf   lower surface of leaf

A   B

AO1 **(a)** Name the parts labelled A and B. [2]

AO1 **(b)** Describe the role of part B in photosynthesis. [3]

AO2 **(c)** Explain the role of structure A in controlling water loss from the leaf. [3]

AO2 **(d)** There is a difference between the numbers of these structures on the upper surface and the lower surface of the leaf.

Suggest how this difference may benefit the plant. [2]

2. In some people, the coronary arteries that carry blood from the heart become narrower as they age.

Narrowed coronary arteries can be treated in several ways, including:

• using a stent to hold the artery open,

• inflating a tiny balloon inside the artery (it does not block the artery).

Three groups of researchers, A, B and C, carried out studies to find out whether using a stent or using a balloon was better.

Each group of researchers measured the percentage of patients whose arteries narrowed again, after treatment with each method.

Their results are shown in the bar chart.

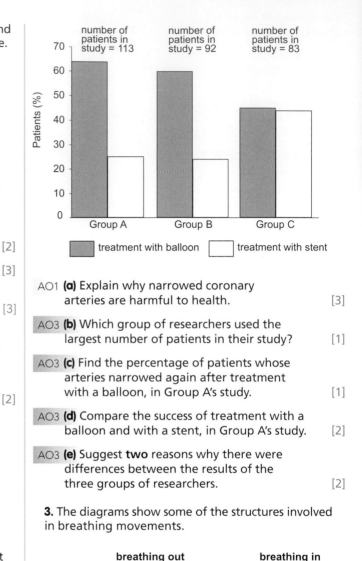

AO1 **(a)** Explain why narrowed coronary arteries are harmful to health. [3]

AO3 **(b)** Which group of researchers used the largest number of patients in their study? [1]

AO3 **(c)** Find the percentage of patients whose arteries narrowed again after treatment with a balloon, in Group A's study. [1]

AO3 **(d)** Compare the success of treatment with a balloon and with a stent, in Group A's study. [2]

AO3 **(e)** Suggest **two** reasons why there were differences between the results of the three groups of researchers. [2]

3. The diagrams show some of the structures involved in breathing movements.

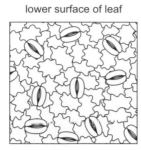

AO1 **(a)** Describe how each of the following have changed between breathing out and breathing in.

• the ribcage

• the diaphragm [2]

AO2 **(b)** Explain how the changes that you have described help to make air flow into the lungs. [3]

AO1   recall the science   AO2   apply your knowledge   AO3   evaluate and analyse the evidence

✳ WORKED EXAMPLE – Foundation tier

The diagram shows a section through an alveolus and a blood capillary in the human lungs.

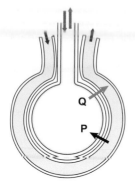

(a) Arrow **P** shows the direction in which oxygen diffuses from the alveolus into the blood capillary.

Explain what causes the oxygen to diffuse in this direction. [2]

It moves down its concentration gradient.

(b) Name the gas that diffuses in the direction of arrow **Q**. [1]

Carbon dioxide.

(c) State **two** features of the alveoli in the lungs that help gases to be exchanged easily and quickly between the alveoli and the blood. [2]

They are very thin. They are very small, so there are a lot of them and they make a large surface area.

How to raise your grade!
Take note of these comments – they will help you to raise your grade.

Notice that this is an 'Explain' question. This means that you need to say why or how something happens.

This answer is correct. However, there were two marks for the question and this answer gives only one item. The candidate has not said enough to obtain all the marks available.

The candidate could have explained what causes the concentration gradient. There is not much oxygen in the blood inside the capillaries because it has come from the body cells that have been using oxygen. There is a lot of oxygen in the alveoli because fresh air is always being breathed into the lungs.

This answer receives one mark.

That is fine, and receives the mark. There is no need to say any more for a 'Name' question.

The candidate has found two different points to make, and has made each of them clearly.

There is no need to explain anything, as the question does not ask for this. However, the candidate should have made it clear that it is the walls of the alveoli rather than the alveoli themselves that are very thin. Also, the key point in the second part of the answer is 'large surface area'. Nonetheless, this answer receives both marks.

1. Some professional cyclists have used a procedure called blood doping before taking part in competitions. Anyone who is found to have done this is disqualified from competing.

Blood doping involves taking blood from the cyclist's body. Some of the liquid is removed from the blood. The blood is stored at a low temperature.

The cyclist's body naturally makes more blood to replace the blood that has been removed.

About 24 hours before the competition, the stored blood is put back into the cyclist's body.

The table shows how this affects the cyclist's blood and their ability to exercise.

When tested	Concentration of haemoglobin in the blood (g/cm³)	Time the cyclist can run on a treadmill at top speed (s)
before the saved blood is transfused	13.6	790
after the saved blood is transfused	17.5	920

AO2 **(a)** Suggest why the blood is stored at a low temperature after it is removed from the cyclist's body. [1]

AO2 **(b)** Using the information above, and your own knowledge about the composition of blood, explain how blood doping affects the concentration of haemoglobin in the blood. [3]

AO3 **(c)** Using the information in the table, suggest how blood doping can help a cyclist to win a race. [3]

2. (a) The photograph shows villi in the digestive system of a bat.

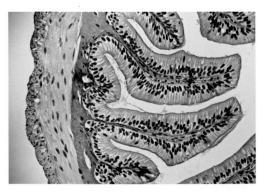

AO1 **(i)** Name the part of the digestive system in which villi are found. [1]

AO1 **(ii)** Describe the function of villi. [3]

AO2 **(iii)** Explain how **two** features of villi, visible in the photograph, adapt them for their function. [2]

(b) Different species of bats eat different diets.

- The vampire bat, *Desmodus rotundus,* feeds entirely on blood.

- *Sturnia lilium* is a fruit bat that eats ripening fruit.

- *Glossophaga soricina* feeds on nectar from flowers.

AO2 **(i)** Suggest which bat has the diet containing the most cellulose. [1]

AO2 **(ii)** Suggest which bat has the diet containing the most protein. [1]

AO2 **(iii)** Suggest which bat has the diet that needs the least digestion. [1]

(c) Researchers put forward the hypothesis that the length of the villi in the intestines of a bat is related to its diet. They measured the villi in these three species of bats. The bar chart shows their results.

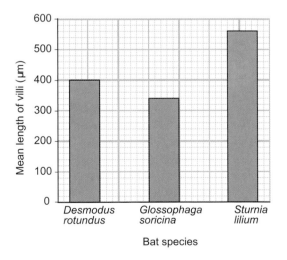

AO3 **(i)** With reference to the bar chart, compare the length of the villi in these three species of bats. [2]

AO3 **(ii)** Discuss the extent to which these results support the researcher's hypothesis. [3]

AO1   recall the science   AO2   apply your knowledge   AO3   evaluate and analyse the evidence

✳ WORKED EXAMPLE – Higher tier

Aphids are insects that feed on plants. They have needle-like mouthparts that pierce the plant stem. They suck out fluid from the plant's phloem tissue.

(a) Explain why the contents of phloem tissue provide a better food source for aphids than the contents of xylem tissue.　　[3]

Phloem tissue contains sugars. This provides nutrients for aphids, because sugars provide energy. Xylem tissue only carries water and mineral ions, and these wouldn't give aphids any energy.

(b) Aphids can be serious pests. Gardeners may spray plants with insecticides.

Contact insecticides kill insects when the spray hits the surface of the insect. Systemic insecticides are absorbed into the plant and transported in the phloem.

Explain why systemic insecticides are likely to kill more aphids than contact insecticides.　　[2]

If you sprayed a plant with a contact insecticide, you might miss some of the aphids. But a systemic insecticide would get carried all over the plant in the phloem. The aphids are sucking from the phloem, so they would suck up the insecticide and be killed.

(c) In humans, oxygen is transported to the body tissues in the blood. In plants, however, neither phloem tissue nor xylem tissue transport oxygen.

Suggest why plants do not need to transport oxygen in their phloem tissue or xylem tissue. You should use the words **surface area** and **diffusion** in your answer.　　[4]

In humans, cells are respiring quite a lot. They need oxygen for respiration, which provides energy to cells. But plants don't move around so they don't need very much energy, so they don't respire so much and don't need so much oxygen.

Also, plants are quite thin and branching. They have a large surface area. Oxygen can get to all parts of the plant by diffusion, because none of the plant cells are far from the surface. In a human, it would take much too long for oxygen to diffuse from the lungs to all the different parts of the body.

How to raise your grade!

Take note of these comments – they will help you to raise your grade.

This is a good answer. The candidate has made a clear statement about what phloem contains, and then explains why this makes it nutritious for aphids. Lastly, they compare this with the contents of xylem. The question asked for a comparison.

The answer receives all three marks.

There is quite a lot of information in this question. Examiners usually do not give information unless you need to use it – read it carefully.

This answer is excellent. The candidate has linked the information about how aphids feed (given earlier in the question) with this new information. Making links like this can raise your answer from good to excellent. This answer receives two marks.

This is a good suggestion. The candidate has thought about why organisms need oxygen, and then suggested why plants do not need as much oxygen as mammals.

Here is another good suggestion. The candidate has thought about gaseous exchange surfaces and diffusion.

The candidate has planned the answer well and receives all four marks.

Biology B3.3–3.4

What you should know

Removal of waste

All organisms produce waste products as a result of the chemical reactions that take place in their cells.

The removal of these waste products is called excretion.

In humans, toxic waste materials are removed from the body by the kidneys.

- Name the liquid produced by human kidneys, in which toxic waste is removed.

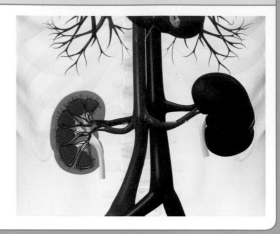

Control in the human body

The water content, ion content, temperature and blood sugar content of the human body are controlled.

Receptors detect changes.

Nerves and hormones allow communication between receptors and effectors, which take action.

- Which communication system works more quickly – nerves or hormones?

Energy and biomass in food chains

Green plants transfer energy from sunlight in photosynthesis.

Food chains show how energy is transferred from one organism to another by feeding.

The amounts of energy transferred are reduced at each step in a food chain.

Photosynthesis removes carbon dioxide from the air. Some of the carbon is used to make new biomass, which contains energy.

- What are the producers in every food chain?

You will find out

Homeostasis: Removal of waste and water control

> Carbon dioxide and urea are produced in the body and must be excreted.

> Urea is made in the liver from excess amino acids.

> Urea is removed by the kidneys and excreted in urine.

> Kidney failure can be treated by dialysis or a transplant.

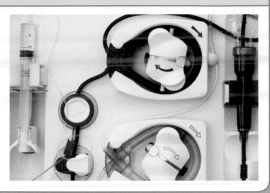

Homeostasis: Temperature and sugar control

> In the human body, temperature, water content and blood glucose concentration are all kept fairly constant.

> Body temperature and water content are monitored by the brain. Skin and muscles help to raise or lower body temperature. Kidneys help to maintain the correct amount of water in the body.

> Blood glucose concentration is monitored in the pancreas. The pancreas secretes insulin and glucagon when blood sugar is too high or too low. The liver responds by using or producing glucose.

Humans and their environment

> Untreated sewage, sulfur dioxide, pesticides and herbicides can harm organisms and their environment.

> Large-scale deforestation can reduce biodiversity and reduce the rate at which carbon dioxide is removed from the atmosphere.

> Rice fields and cattle produce methane. Methane traps heat in the atmosphere and increases global temperatures.

> Biofuels are produced from living organisms.

> Food can be produced more efficiently by eating plants rather than animals, and by keeping animals warm.

> Fish stocks are dwindling and must be conserved.

Waste and water control

You will find out:

> carbon dioxide and urea have to be removed from the body
> the water and ion content of the body must be controlled
> how kidneys remove urea, excess ions and water

Kangaroo rats

Kangaroo rats live in deserts. They almost never drink water, obtaining all that they need from the seeds that they eat. Their kidneys are superbly efficient at conserving water, producing urine that is 17 times more concentrated than the kangaroo rat's blood.

FIGURE 1: A kangaroo rat's efficient kidneys enable it to live in places where humans would quickly die of dehydration.

Waste

Waste products

Chemical reactions in body cells produce waste products. These waste products must be removed from the body.

Carbon dioxide is a harmful waste product of respiration.

> It is produced by all body cells.

> It is removed from the lungs when we breathe out.

Urea is a harmful waste product produced in the liver.

> The liver makes urea from amino acids, if we have more of them in the body than we need.

> The urea is transported to the kidneys in the blood.

> The kidneys make **urine**, which contains urea.

> The urine is passed out of the body.

Keeping the balance

It is important to keep the right amount of water and ions (such as sodium ions or chloride ions) in the body. The kidneys remove just the right amount of these substances from the blood.

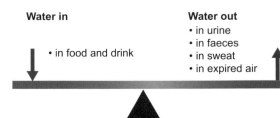

Water in
• in food and drink

Water out
• in urine
• in faeces
• in sweat
• in expired air

FIGURE 2: Water balance.

⊙ QUESTIONS

1 Copy and complete this table.

Waste substance	Where it is made	How it is made	How it is removed
	in all body cells		by the lungs
urea		from excess amino acids	

Control

How kidneys produce urine

Blood containing water, waste products, blood cells and useful nutrients – such as glucose – flows to the kidneys. The kidneys extract waste products and spare water, and use them to make urine. The kidneys make sure that glucose and other useful substances stay in the body.

Q ... water balance homeostasis AND urea

1. The kidneys **filter** the blood. Blood cells cannot pass through the filter. Small molecules – which include water, glucose and urea – and ions do pass through the filter. These molecules go into the tubes inside the kidney.

2. Next, the kidneys put back any useful substances into the blood. This includes all of the glucose, any of the ions that the body needs, and some of the water. This process is called **reabsorption**.

3. What is left in the kidney is urea, excess ions and excess water. This mixture is urine. Urine flows from the kidneys to the bladder, where it is stored until it is removed from the body.

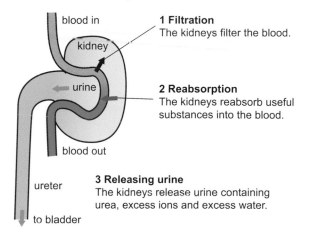

1 Filtration
The kidneys filter the blood.

2 Reabsorption
The kidneys reabsorb useful substances into the blood.

3 Releasing urine
The kidneys release urine containing urea, excess ions and excess water.

FIGURE 3: How kidneys make urine.

Homeostasis

The kidneys help to keep the water and ion content of the body fairly constant. This helps body cells to work well. The surroundings of cells – the **internal environment** of the body – always contains a similar concentration of ions and a similar amount of water.

There are other aspects of the internal environment that are kept fairly constant. These include temperature and glucose concentration.

Keeping the internal environment constant is called **homeostasis**.

QUESTIONS

2 Urine made by a healthy kidney does not contain any blood cells. Explain why.

3 If a person has been doing a lot of exercise on a hot day, they will probably produce a smaller volume of urine than if they had been resting. Assuming that they drink the same amount, suggest an explanation for this smaller volume.

4 Suggest why it is important that the water and ion content surrounding the body cells is kept at just the right level.

In as food, out in urine

Surveys show that many people think there is a direct connection between their digestive system and their kidneys. They think that waste liquid from the digestive system just flows out as urine.

This is not correct. Figure 4 shows how any excess protein in a person's food can end up as urea in urine.

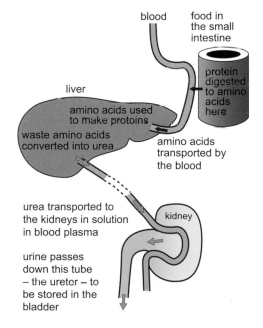

FIGURE 4: How urea is produced and removed from the body.

QUESTIONS

5 The table shows the quantities of various substances found in urine and in blood plasma. For each substance, explain the reasons for the differences in the percentages.

Substance	In blood plasma (%)	In urine (%)
water	92	95
glucose	0.1	0
salt (NaCl)	0.4	0.6
urea	0.03	2

Treating kidney failure

You will find out:

> how kidney dialysis works
> how kidney transplants are done
> some of the advantages and disadvantages of dialysis and transplants

Kidney damage

The MRI scan, in Figure 1, was taken to find out what was wrong with a man's kidneys. The kidneys are the two green objects, on either side of the spine (which looks white). The kidney on the right is enlarged and has a brown patch in it. This is an abscess, caused by an infection.

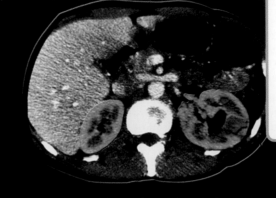

FIGURE 1: Even if the abscess damages one kidney, the person will survive perfectly well with the other healthy kidney.

Kidney failure

Kidneys can fail for many reasons. For example, an infection may damage the kidney cells so badly that they cannot work any more. Once both kidneys stop working, then waste substances build up in the blood, and the water balance is no longer kept steady.

There are two treatments for kidney failure – **dialysis** and **transplant**.

Dialysis

A dialysis machine acts like a substitute kidney.

> The patient's blood is passed through it.

> Waste substances are removed from the blood.

> The machine restores the concentration of dissolved substances in the blood to normal levels.

> The blood is allowed to flow back into the patient's body.

Dialysis has to be carried out every few days.

Transplant

A healthy kidney from one person can be transplanted into a person with kidney failure. They only need one kidney – it can do all the work needed to keep the body healthy.

The person the kidney is taken from is the **donor**. The person receiving the kidney is the **recipient**.

The donor is often someone who has just died, perhaps in an accident. However, it is also possible for a living person to donate a kidney. Close relatives of the patient sometimes offer to do this.

One major problem with transplants is that the recipient's body may **reject** the donated kidney. Precautions are taken to avoid this rejection.

QUESTIONS

1 Explain why it is important to treat kidney failure quickly.

2 State two ways in which kidney failure is treated.

3 Suggest why many patients would prefer to have a kidney transplant, rather than rely on dialysis.

How dialysis works

Kidney dialysis

The patient's blood flows between partially permeable membranes. These membranes separate the blood and the dialysis fluid.

The tubes carrying the dialysis fluid and the blood are very long and narrow.

Fresh dialysis fluid does not contain urea. Urea diffuses out of the blood into the dialysis fluid.

Dialysis fluid contains the same concentration of useful substances – such as glucose – as the blood, but no urea.

Advantages and disadvantages of dialysis or a transplant

Most patients with kidney failure are kept healthy with dialysis until a transplant is available for them.

Most patients want to have a transplant, because it will free them from having to be hooked up to a dialysis machine for hours on end. However, even transplants have drawbacks.

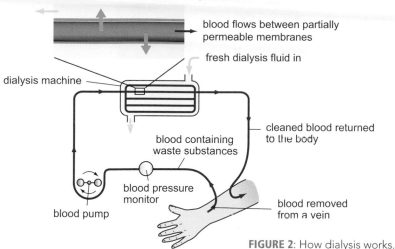

urea diffuses out of the blood into the dialysis fluid

blood flows between partially permeable membranes

fresh dialysis fluid in

dialysis machine

cleaned blood returned to the body

blood containing waste substances

blood pressure monitor

blood pump

blood removed from a vein

FIGURE 2: How dialysis works.

QUESTIONS

4 Explain how these features of a dialysis machine help it to work effectively: (a) the tubes (b) the partially permeable membrane.

5 People with kidney failure must not eat salty foods, except while they are attached to a dialysis machine. Suggest why.

Treatment	dialysis	transplant
Advantages	> Dialysis machines are more widely available than kidneys for transplant.	> If the transplant works, the person can live a normal life.
Disadvantages	> The patient has to be treated for several hours, several times a week. This restricts activities such as going on holiday. > Many people feel ill while they are having dialysis. > Dialysis does not cure the problem – it just keeps the patient alive.	> There are not enough donor kidneys available for all the people who need one. > The donor kidney may be rejected by the person's immune system. The person has to take drugs to stop this rejection, which can make it more difficult for them to fight off infections.

More about rejection

The body's immune system has evolved to attack and destroy invading pathogens. Unfortunately, it treats cells in an organ that has come from another person's body in the same way as it would treat harmful bacteria.

On the surface of every cell there are proteins called **antigens**. Antigens on the cells in the donor kidney will be different from the antigens on the recipient's cells. The recipient's white blood cells recognise this, and make antibodies that stick to the antigens. This destroys the cells in the donor kidney. The kidney has been 'rejected'.

There are two main ways in which this can be avoided.

> Donor kidneys are tested to check whether they have a similar 'tissue type' to the recipient, meaning that their antigens are similar. This is why a kidney from a living, close relative may be used.

> The recipient is given drugs that make the immune system work less effectively. These are **immunosuppressant** drugs. The recipient will have to take them all their life.

QUESTIONS

6 Should it be assumed that everyone is willing for their kidneys to be used if they die suddenly, even if they are not carrying a donor card? Discuss whether or not this is a good idea.

Temperature control

You will find out:

> human body temperature is kept fairly constant

> body temperature is monitored by the brain

> sweat glands and blood vessels in the skin change the rate of heat loss

A mouthful of maggots

Fishermen often use live maggots (fly larvae) as bait. The maggot is more likely to attract a fish if it is wriggling. Warm maggots move faster than cold maggots, so some fishermen put the maggot into their mouth for a while, to warm it up before they put it onto the hook.

FIGURE 1: Maggots cannot control their temperature – on a cold day their bodies are cold and slow moving.

Maintaining a constant temperature

We are mammals and – like all mammals, and also birds – we control the temperature inside our bodies. Normal human body temperature is 37 °C.

Table 1 shows what happens when your body temperature deviates from normal.

TABLE 1: Body temperature and its effect.

Body temperature (°C)	What happens to a person
38–40	a fever and feel very ill
37	at the right temperature
36	feel cold and a bit shivery
35	shiver almost uncontrollably
32	lose feeling and may even stop feeling cold
30	become unconscious
27	stop breathing

thermoregulatory centre

If the blood is too hot, the thermoregulatory centre switches on cooling mechanisms.

If the blood is too cold, the thermoregulatory centre switches on warming mechanisms.

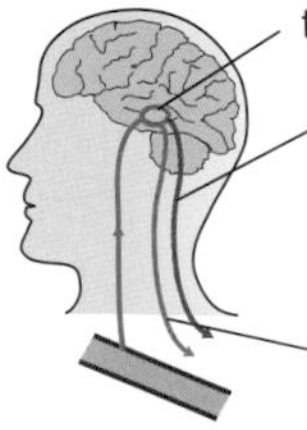

FIGURE 2: The brain helps to control body temperature.

Monitoring and controlling body temperature

The **thermoregulatory centre** in the brain checks the temperature of the blood flowing through it. There are also **temperature receptors** in the skin. These record the temperature of the surface of the skin.

If the temperature of the blood is too high, the brain sends impulses along nerves to **sweat glands** in the skin. This will cause them to secrete more sweat, which cools down the body.

QUESTIONS

1 Where is the body's thermoregulatory centre?

2 State two places in the body where temperature is constantly monitored.

3 Controlling body temperature is part of the way in which the body's internal environment is kept constant. What is this called?

The skin and temperature control (Higher tier)

The skin is the largest organ of the human body. It has a very important role to play in helping to regulate the temperature inside the body.

Temperature receptors in the skin

Temperature receptors in the surface of the skin monitor the skin temperature. When you say 'I feel cold', you are really responding to the information coming from these skin receptors. Your **core body temperature** (the temperature deep inside your body) is probably still 37 °C.

Sweat glands and skin capillaries

The sweat glands and blood vessels in the skin can change the rate at which heat is lost from the body.

When you are too hot:

> arterioles (small arteries) supplying blood to the capillaries near the surface of the skin dilate (widen). More blood flows close to the skin surface. Energy is lost from the blood by radiation

> sweat glands secrete more sweat, which lies on the surface of the skin. The water in the sweat evaporates. This takes heat from the skin, cooling it.

When you are too cold:

> arterioles supplying the skin capillaries constrict (become narrower). Less blood flows close to the skin surface, so less heat is lost to the air.

> muscles in various parts of the body contract very quickly. This is shivering. The energy for the muscle contraction is released by respiration. This also produces heat energy, which warms the body.

Remember

The skin capillaries do not move when you are hot – they just carry more blood.

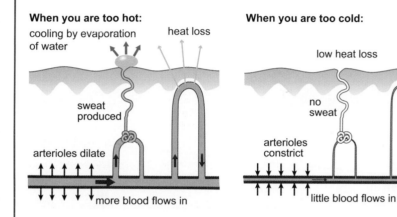

FIGURE 3: How the skin regulates heat loss.

QUESTIONS

4 Explain how sweating helps to cool a person.

5 Explain how constriction of arterioles helps to reduce heat loss.

Advantages of controlling body temperature

Most animals cannot control their body temperature. Their temperature is the same as their environment.

A constant warm body temperature helps cells to work at their best.

> If cell temperature drops much below 37 °C, then the reactions in them slow down. The activity of the whole animal is also slowed down.

> If the temperature goes much higher than 37 °C, then enzymes are denatured, again slowing down or even stopping the reactions.

QUESTIONS

6 Suggest why mammals and birds are able to live in the Arctic, while reptiles and amphibians cannot.

7 Mammals that live in very cold water, such as whales and walruses, are often very large and have thick layers of fat under their skin. Suggest how these two features can help such mammals to control their core body temperature.

FIGURE 4: Walruses can survive in water so cold that it would kill a human within minutes.

Controlling blood glucose

You will find out:

> the pancreas monitors and controls blood glucose level

> insulin reduces blood glucose level

> glucagon increases blood glucose level

> how Type 1 diabetes is caused and controlled

A cure for diabetes?

Increasing numbers of people are developing diabetes, an illness in which the control of blood glucose goes wrong. It is often caused by cells in the pancreas not working properly. At the moment, diabetes can only be controlled, not cured. It is hoped that, one day, stem cells could be used to replace faulty pancreas cells, and thus provide a permanent cure.

Did you know?

Almost 2.5 million people in the UK have diabetes – and the numbers keep increasing.

FIGURE1: Many people with diabetes need to test their blood glucose level regularly.

Blood glucose

Blood glucose levels

Glucose is transported around the body dissolved in blood plasma. It is the main fuel for body cells.

Cells take glucose and combine it with oxygen in respiration. This releases energy that the cell can use for its activities.

If blood glucose level drops too low, then cells may not get enough energy. Brain cells are often the first to be affected, so the person may feel faint or even lose consciousness.

If blood glucose levels rise too high, then water is drawn out of body cells by osmosis. This can do permanent harm.

Insulin and diabetes

When blood glucose level rises, the **pancreas** detects this. Cells in the pancreas respond by secreting a hormone called **insulin**. Insulin increases the rate at which glucose moves out of the blood and into cells.

In some people, the pancreas does not produce enough insulin. Their blood glucose level can therefore rise too high. This is **Type 1 diabetes**.

Type 1 diabetes cannot be cured.
It can be controlled by:

> taking care with diet – eating small amounts at regular intervals

> balancing diet with exercise – if you do a lot of exercise you can eat more carbohydrate

> injecting insulin to bring blood glucose levels down.

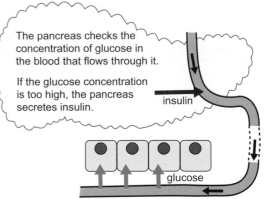

The pancreas checks the concentration of glucose in the blood that flows through it.

If the glucose concentration is too high, the pancreas secretes insulin.

insulin

glucose

Insulin makes body cells take in glucose from the blood.

FIGURE 2: How the pancreas reduces high blood glucose levels.

QUESTIONS

1 Name the body organ that monitors blood glucose concentration.

2 Name the body organ that secretes a hormone to control blood glucose concentration.

3 Under what conditions is insulin secreted?

4 What does insulin do?

Controlling Type 1 diabetes

In Type 1 diabetes, not enough insulin is secreted. A person with Type 1 diabetes has to learn how to control their blood glucose through their behaviour, rather than being able to rely on their body to do it automatically.

Diet

A person with Type 1 diabetes must take care not to eat too much carbohydrate at once. This would increase their blood glucose level. They also must make sure that they do not go too long between meals containing carbohydrate. This could cause their blood glucose level to fall too low. They need to eat carbohydrate little and often.

Injecting insulin

Most people with Type 1 diabetes have to inject insulin each day. They will probably need to test their blood glucose level to check how much insulin they need.

Exercise

Many people with Type 1 diabetes lead very active lives. They do need to be aware that doing exercise will use up glucose, so they need to eat the right amount of carbohydrate to compensate for this.

They may need to inject less insulin on days when they use up a lot of energy.

FIGURE 3: Having diabetes does not have to stop you doing active things such as backpacking trips.

QUESTIONS

5 The hormone insulin is a protein. Swallowing an insulin pill will have no effect. Suggest why.

6 Explain why a person may have to inject less insulin on days when they use up a lot of energy.

Insulin and glucagon (Higher tier)

When insulin is secreted, it makes cells take up more glucose from the blood. Insulin also makes liver cells change glucose into a substance called **glycogen**.

Glycogen has molecules made of hundreds of glucose molecules linked together in a long chain. It is insoluble, and can be stored easily, in liver cells.

If blood glucose levels drop too low, then the pancreas stops secreting insulin and secretes another hormone, called **glucagon**. Glucagon makes liver cells break down some of the glycogen they have stored, turning it back into glucose. The glucose passes out of the liver cells and into the blood.

If a person's blood glucose levels rise very high, glucose may be excreted in their urine. In the past, a person would test their urine for glucose, but nowadays they normally test their blood.

Remember

Glucagon is a hormone. Glycogen is a carbohydrate stored in animal cells.

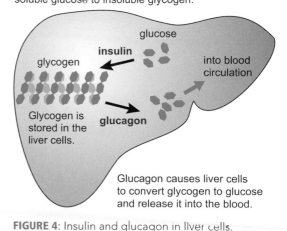

Insulin causes liver cells to convert soluble glucose to insoluble glycogen.

Glycogen is stored in the liver cells.

Glucagon causes liver cells to convert glycogen to glucose and release it into the blood.

FIGURE 4: Insulin and glucagon in liver cells.

QUESTIONS

7 Testing blood instead of urine allows someone with diabetes to control their blood glucose level more successfully. Suggest why this is more successful.

Preparing for assessment: Planning an investigation

To achieve a good grade in science, you not only have to know and understand scientific ideas, but you need to be able to apply them to other situations and investigations. These tasks will support you in developing these skills.

✸ How does temperature affect the rate of dialysis?

Vicki decided to make a model of a dialysis machine, to help her to investigate how temperature affects the rate of dialysis.

Method

Vicki's teacher showed her how she could make a simple model of a dialysis machine (Figure 1).

> The contents of the Visking tubing represent the blood of the patient that flows through the machine.

> The contents of the beaker represent the dialysis fluid that also flows through the machine.

> The blood and the dialysis fluid are separated by the dialysis membrane.

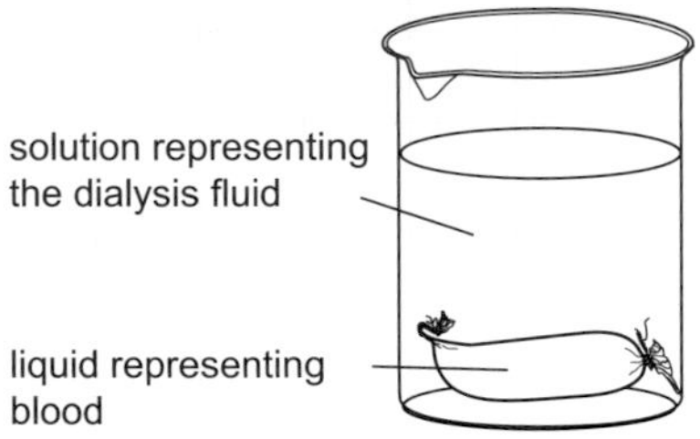

solution representing
the dialysis fluid

liquid representing
blood

FIGURE 1: Model of a dialysis machine.

Various materials may be used to make the 'blood' and the 'dialysis fluid'. For example, Vicki's teacher suggested using yellow food dye, which can pass through the holes in Visking tubing, to represent urea.

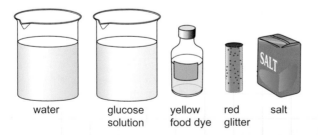

water   glucose solution   yellow food dye   red glitter   salt

FIGURE 2: Representing blood and dialysis fluid.

✳ Planning

1. Write down a hypothesis that Vicki could test.

2. Suggest how Vicki could use the substances shown in Figure 2 to make the two liquids to go into her model of a dialysis machine. Explain your suggestions.

3. Predict whether each of these substances used to make 'blood' would move through the Visking tubing.

4. What will be the independent variable in Vicki's investigation? Suggest how Vicki can alter this variable. What range should she use?

5. Vicki decided to measure the rate at which the yellow food dye diffused out of the 'blood' and into the 'dialysis fluid' as her dependent variable.

(a) Do you think that this is a good choice of what to measure?

(b) Suggest how Vicki could measure this variable.

6. Suggest two variables that Vicki should keep constant in her investigation.

> It should make a clear statement about how temperature affects the rate of dialysis.

> Think about what blood contains. You cannot make anything that is exactly like real blood, but you could get quite close. If you think that something should go into both the 'blood' and into the 'dialysis fluid', should their concentrations be the same, or different?

> Firstly, decide whether this would be a valid measurement to test Vicki's hypothesis – would it actually test what she wants to test? Secondly, think about whether it would be possible to measure it, and to do so with reasonable precision.

> How might Vicki use a set of 'standard' solutions, against which she could compare the colour of the liquid? How should she make up these standard solutions?

✳ Assessing and managing risks

7. Assess the risks involved with the methods that you have described in your answers. Explain what you would do to reduce these risks.

✳ Collecting primary data

8. Construct a results table in which you would record your results.

> The headings of your table will depend on what you think Vicki will be measuring.

✳ Connections

How Science Works

• Plan practical ways to develop and test scientific ideas

• Assess and manage risks when carrying out practical work

• Collect primary and secondary data

Science ideas

B3.3.1 Removal of waste and water control

Waste from human activities

You will find out:

> the growing human population is producing more and more waste

> waste can pollute air, water and land

> dumping waste, building, farming and quarrying reduce the amount of land where wildlife can live

Mountains of waste

We produce enormous amounts of waste. Much of it could still be useful. Often, it is too much trouble to pass on those things that we see as rubbish but that someone else could use. Not only is that wasteful, but disposing of this rubbish causes damage to the environment.

FIGURE 1: These people make a living by searching for food and other useable waste on a rubbish tip.

The waste problem

Population growth

The human population on Earth is growing. There are signs that this growth is slowing down, but numbers will continue to increase for many years to come.

Problems with waste disposal

The more of us that there are, the more waste we produce. Many people now have a higher standard of living than in the past, which means they have more things to throw away. Unless this waste is handled carefully, it may seriously damage the environment. Waste can cause pollution, harming plants and animals – including ourselves.

Less land for wildlife

An increasing human population means a decrease in the amount of land available for other animals and plants.

Land is used for:

> **building** houses, roads, schools, offices, hospitals and factories

> **quarrying** to obtain materials from the ground, such as coal, metal ores, salt and stone for building

> **farming** to produce food.

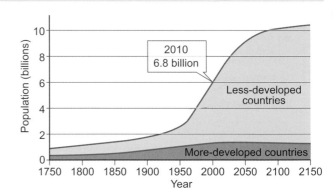

FIGURE 2: It is probable that human population growth will continue into the next century, although it is slowing down.

FIGURE 3: This field near Rotherham was a landfill site – the rubbish is decaying under the ground.

QUESTIONS

1 Give two reasons why the amount of waste that humans produce is increasing.

2 List three ways that the human population uses land.

Waste and pollution

Dumping waste on land can cause land pollution. Toxic chemicals such as pesticides or herbicides can be washed out of the waste and spread to other places. Dumping waste on land also uses up land where wildlife could live.

Untreated sewage, fertilisers and toxic chemicals (perhaps from factories) can also kill plants and animals in water.

Waste gases and smoke can pollute the air. For example, burning fossil fuels can produce sulfur dioxide, which contributes to acid rain.

TABLE 1: Pollution.

Pollutant	Where it comes from	How it causes harm
untreated sewage	houses, factories, runoff from city streets	lowers oxygen concentration in water, killing aquatic animals
fertiliser	runoff from farmland	lowers oxygen concentration in water, killing aquatic animals
toxic chemicals	mostly from factories	can kill organisms on contaminated land or in water
smoke	burning waste or fossil fuels	can cause asthma attacks or respiratory illness
sulfur dioxide	burning fossil fuels containing sulfur (particularly coal)	contributes to acid rain, which kills plants and aquatic animals
pesticides	farming and horticulture – used to kill insects and other pests	can kill animals that are not pests
herbicides	farming and horticulture – used to kill weeds	can kill plants that are not weeds some can harm animals

FIGURE 4: Farmers can test the soil for plant nutrients in different parts of a field, and then use GPS to apply the right amount of fertiliser to each area. How can this help to reduce pollution?

QUESTIONS

3 Many people confuse fertilisers, pesticides and herbicides. Describe the differences between them.

4 Expensive fertilisers are wasted if washed off the land and into rivers. Suggest how the following could help to reduce water pollution.

(i) Farmers do not apply fertilisers just before it rains.

(ii) Farmers apply fertilisers only when crops are growing.

A balancing act

It is not easy to solve all the problems caused by the increasing human population. Advantages and disadvantages have to be weighed up before making decisions about the best thing to do.

For example, in the 1990s there were often traffic jams on the M6 motorway through Birmingham. Vehicles took longer to make their journeys, causing more pollution from their exhaust gases.

A toll road was built to take some of the traffic from the M6. This has made journeys faster, which could reduce the amount of pollution. On the other hand, it may result in more people making more journeys. In addition, the road has taken up a lot of land.

QUESTIONS

5 Discuss whether or not building new roads is good or bad for the environment.

FIGURE 5: The construction of the M6 toll road used land that might otherwise have been available for farming or for wildlife.

Deforestation

You will find out:
> deforestation is happening on a large scale in tropical areas
>loss of forests and peat bogs causes loss of biodiversity and an increase in CO_2 in the atmosphere

No place for the lemurs?

The island of Madagascar, off the southeast coast of Africa, has many species of animals that live nowhere else in the world. It is the only place that lemurs live naturally. Deforestation is happening there at a frightening speed. Soon, there may be nowhere for the lemurs to live.

Did you know?

At the start of the 20th century, only 5% of Britain was covered by forest. It is now about 12%.

FIGURE 1: Lemurs evolved on Madagascar after it became isolated from Africa.

Reasons for deforestation

People have always used trees. Most trees regrow from the roots after they have been cut down. If woodland is managed carefully, you can continue to harvest timber from it for hundreds of years.

If the roots are destroyed, then the trees die. This is often done by first cutting down the trees and then burning the remains.

Many different species of animals and plants are adapted to live in forests.

> If the forest is destroyed, animals and plants lose their habitat.

> Deforestation reduces the number of different species that can live in an area – it reduces **biodiversity**.

We removed most of the forests in the UK centuries ago. The amount of forest is actually now increasing.

Today, the concern is forest in the tropics. Tropical forests have really high biodiversity. However, the people who live there, such as in Madagascar, may be very poor.

Tropical forests are destroyed:

> to provide more land for farming animals and crops.

> to sell the timber

> to create land for producing biofuels.

FIGURE 2: This once-forested hillside in Brazil will now be used for growing crops. The bare soil will easily erode when it rains.

QUESTIONS

1 Explain why simply cutting down trees is not deforestation.

2 What is meant by 'biodiversity'?

3 Explain why deforestation reduces biodiversity.

Carbon dioxide levels

Trees, like all green plants, take in carbon dioxide (CO_2) from the air and use it in photosynthesis. The carbon from the CO_2 becomes part of the cells in the tree. A tree – especially its wood – contains large amounts of carbon.

If trees are cut down and burned, this carbon goes back into the air in the form of CO_2. Even if the trees are left to rot, the carbon will still go back into the air. This happens while microorganisms cause the tree to decay, and their respiration produces CO_2.

Moreover, if the trees are replaced by buildings or roads, then there are not as many plants to take carbon dioxide out of the air. For all of these reasons, deforestation is contributing to the increase of CO_2 in Earth's atmosphere.

The importance of peat bogs

Peat contains carbon that plants took in when they were photosynthesising.

In a peat bog, plants do not rot fully.

1. The ground stays very wet.

2. Decay microorganisms cannot obtain enough oxygen to respire.

3. The partly-rotted plant remains build up in thick, dark layers.

Farmers cannot grow crops on peat but peat has other uses:

> It makes a good fuel.

> Peat is added to soil in gardens, to improve its water-holding abilities.

Overuse has had consequences:

> It has led to the destruction of many peat bogs.

> As the exposed peat dries out, microorganisms break it down, releasing locked-up carbon as CO_2.

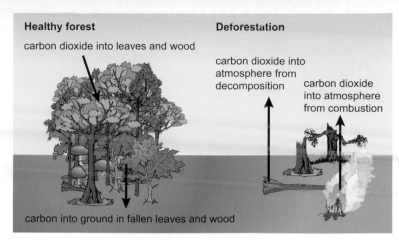

Healthy forest — carbon dioxide into leaves and wood

Deforestation — carbon dioxide into atmosphere from decomposition; carbon dioxide into atmosphere from combustion

carbon into ground in fallen leaves and wood

FIGURE 3: How deforestation can increase CO_2 levels in the atmosphere.

FIGURE 4: A peat bog in Yorkshire.

QUESTIONS

4 Explain why deforestation is contributing to the increase in the concentration of CO_2 in the atmosphere.

5 Gardeners can now buy peat-free compost. Explain how choosing this kind of compost might help the environment.

Deforestation rates

The National Institute for Space Research in Brazil uses satellite images to help to keep track of deforestation. Table 1 shows their estimates of the rate at which forest has been lost in the Amazon region between 2001 and 2009.

QUESTIONS

6 Describe the changes in the rate of deforestation in the Amazon region between 2001 and 2009.

7 Discuss what could be done to decrease the rate of deforestation, in future.

TABLE 1: Loss of forest in the Amazon region.

Year	Rate of loss (km²/year)
2001	18165
2002	21393
2003	25247
2004	27423
2005	18846
2006	14109
2007	11532
2008	12911
2009	7008

Q ... peatbog ... deforestation Amazon Brazil

Global temperature

Algae power

This aeroplane, on display at Farnborough in 2010, is the first that flies on fuel made from algae. The fuel is (energy rich) – less fuel is needed than normal jet fuel, while maintaining the same performance.

FIGURE 1: Using fuel made from algae reduces this aeroplane's 'carbon footprint'.

You will find out:

> increasing levels of carbon dioxide and methane in the atmosphere are contributing to global warming

> global warming may affect the distribution and abundance of species

> oceans can soak up and hold carbon dioxide

> using biofuels helps to reduce overall carbon dioxide emissions

Global warming

Measurements of temperatures on Earth show that the mean global temperature is increasing. This is called **global warming**. There may be several reasons for this temperature rise. However, most scientists believe that human activity is one of these reasons.

Human activities have caused the levels of carbon dioxide (CO_2) and methane (CH_4) in the atmosphere to increase. Both of these gases trap heat, keeping Earth's surface warmer than it would otherwise be.

Effects of global warming

Even a small increase in temperature – just a few degrees – is likely to have many effects.

> Changes in climate – For example, some places would have less rain. Other places would have more extreme weather, such as hurricanes or heavy snowfall.

> Rise in sea level – Water expands when it is heated. Ice will melt in the Arctic and Antarctic, adding more water to the oceans.

> Reduction of biodiversity – Some species might be unable to adapt to a changing climate, and might become extinct.

> Changes in migration patterns – Birds that usually spend the winter in warmer countries might stay in cooler ones.

> Changes in the distribution of species – For example, animals that have not been able to live in Britain, because our climate is too cold, might be able to survive here.

FIGURE 2: The small red-eyed damselfly used to live only in warmer parts of Europe, but is now living in Britain.

QUESTIONS

1 What is global warming?

2 How are humans thought to be contributing to global warming?

3 Outline how global warming might affect the distribution of species.

Did you know?

Global temperatures could rise by 4 °C by 2050 – causing a 0.5 m rise in sea level.

Slowing the CO₂ increase

Natural processes

It would be good to slow down the rate at which carbon dioxide levels in the atmosphere are increasing. Earth's natural processes provide some help.

> Fossil fuels, peat bogs and forests are carbon stores.

> Oceans take some carbon dioxide from the atmosphere and store it. This is called carbon **sequestration**. The carbon dioxide dissolves in the water.

> Algae use it in photosynthesis and animals may use it to help make their shells and skeletons. When these algae and animals die, they sink to the bottom of the ocean, where some of them will become part of rocks. The carbon dioxide is locked up in the rocks.

Biofuels

Fossil fuels were formed million of years ago, from dead plants and microorganisms. When they burn, carbon that was trapped in the fuel is released as carbon dioxide.

Reducing the quantity of fossil fuels burned reduces the amount of carbon dioxide added to the air.

Biofuels are fuels that are made from living organisms. They include:

> **biogas** – a mixture of methane and other gases, produced when microorganisms decay biomass

> **bioethanol** – produced by allowing yeast to ferment sugars obtained from crops such as maize

> **biodiesel** – made from oil obtained from plants such as rapeseed or oil palms.

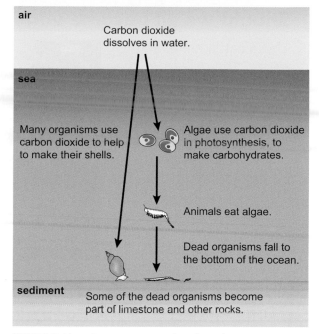

FIGURE 3: How the oceans can absorb and store carbon dioxide.

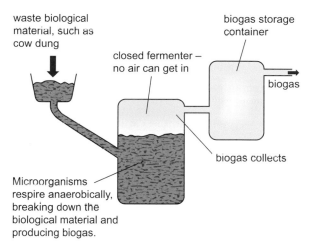

FIGURE 4: A simple biogas generator.

● QUESTIONS

4 Explain the term 'carbon sequestration'.

5 Explain how the oceans may help with carbon sequestration.

6 Suggest how an increase in temperature might affect the rate of production of biogas from a biogas generator. Explain your suggestion.

Making predictions

Scientists use computer programmes to predict how much global temperatures are likely to rise. However, it is impossible to be sure, because:

> all of the factors that are affecting global temperature are not fully understood

> some factors, such as the future emissions of carbon dioxide, cannot be predicted.

● QUESTIONS

7 Discuss how uncertainties about global warming may affect the willingness of individual people, businesses and governments to act to reduce carbon dioxide emissions.

Food production

Should we all become vegetarians?

These cattle are eating maize. The milk from the cattle will be used to make food, or they might be slaughtered to produce meat. Is this a good use of resources? Some scientists think that we would have a less harmful impact on the environment if we all became vegetarians.

FIGURE 1: If we ate this maize, rather than feeding it to cattle, would it help the environment?

Efficient production

Short food chains

Energy is lost as it passes along food chains. This is why biomass decreases as you go along the chain.

For example, maize can be eaten by animals and cattle.

If humans were all vegetarians, we would always be at the end of a short food chain:

maize plants → humans

If humans eat meat or drink milk, the food chain is longer:

maize plants → cattle → humans

By reducing the number of steps in the food chain, the **efficiency** of food production is increased. In theory, more people could be supported on a given area of land.

However, animals can often eat food that humans cannot eat directly. For example, cows can eat and digest grass. Humans cannot digest leaves very well. Humans could not obtain many nutrients from eating leaves of grass.

Food miles

The food on sale in a supermarket may have travelled hundreds or even thousands of miles to arrive there. Supermarket chains have central distribution points.

A carrot might have to be transported many miles to a distribution point before going back to a supermarket near the field where it was grown. This can increase the cost of the food. It also increases carbon dioxide emissions, from the lorries transporting it.

Today, people are encouraged to buy locally-produced food, perhaps from farm shops or local markets.

FIGURE 2: It would not be possible to grow crops on this thin soil. Only grass will grow, which can be grazed by sheep.

QUESTIONS

1 Explain why it is more energy-efficient to eat plants rather than animals.

2 Allowing cows to eat the grass in a field is better than feeding the grass to humans. Explain why.

Q ... food chains cycles ... food miles carbon dioxide emissions

Reducing energy losses

Farmers think hard about how to reduce energy losses along food chains. This can reduce the amount they have to pay for food for their livestock and increase the money they receive for the milk or meat that they sell.

Mammals and birds use energy to keep their bodies at a constant temperature. If a farmer keeps his animals warm, the animals need less food because they use less energy to keep warm.

Animals use up a lot of energy moving around. If they are kept in a small area where they cannot move very much, then more of the energy in their food can go into making more meat, milk or eggs. Less of the energy in the food is wasted.

FIGURE 3: Keeping these piglets warm increases their growth rate. Can you explain how?

FIGURE 4: Keeping hens like this increases egg production. Can you explain why?

QUESTIONS

3 Discuss how farmers can balance the need to produce food cheaply, but still make sure that their animals have a good quality of life.

Food supplies in the developing world

In many parts of the world, people struggle to grow enough food on the land to feed themselves. They may not be able to afford imported food.

For people in many developing countries, being vegetarian is not an option.

> Land may be very dry, or too poor in mineral ions to be able to grow crops that humans can eat. There may be animals – such as goats or sheep – that can survive on the sparse vegetation.

> There may be only a few different crops that can grow on the land.

> People need the extra protein and vitamins that they can obtain only from meat or milk.

> Animals survive through the winter when crops may all be gone, providing food when no plant food is available.

QUESTIONS

4 For each of the following statements, put forward at least one argument for and one against.

(i) People who live as subsistence farmers (that is, they grow all their own food) should be encouraged not to keep animals.

(ii) The UK should not import food from developing countries such as Kenya.

FIGURE 5: This buffalo is ploughing in a rice field. How can using animals in this way benefit the environment?

Fish and fungi

A dangerous job

Deep-sea fishermen lead a hard life. It is cold, wet and smelly and can be dangerous. It is not always easy to make a good living, because fish populations have been decreasing.

FIGURE 1: Why is there a limit on the numbers of fish that are caught?

Fishing and fermenting

Conserving fish stocks

If we take fish from the sea faster than they can be replaced by breeding, there will be fewer fish. This has been happening all around the world.

If numbers fall very low, then some species may disappear completely from some areas. They might even become completely extinct.

It is not easy to reduce the number of fish that are caught commercially. Fishermen need to make a living and people want to buy cheap fish to eat.

Managing fish stocks so that we can continue to harvest them for years to come is an example of **sustainable food production**.

Stocks are being managed in the following ways.

> Banning the use of nets with small holes – Larger holes mean that only larger, adult fish will be trapped. The smaller ones can escape. They will have time to grow into adults and reproduce.

> Imposing quotas – This means allocating a particular quantity of fish, of a particular species, that can be caught each year. Europe has done this, and shares the quotas out between different countries.

Other sources of food

Farming is not the only way to produce food. Long ago, before farming was invented, people used to harvest all their food from the wild. It is still done today, when wild fish is harvested from the sea.

Another way of producing food is to grow microorganisms in fermenters. For example, the fungus *Fusarium* is used to make a food called **mycoprotein**. You can buy this in supermarkets as *Quorn®*.

FIGURE 2: This is an adult cod fish. How can we make sure there are still cod fish around in years to come?

QUESTIONS

1 What is mycoprotein?

2 Why is it important to conserve fish stocks?

3 Describe two ways that could help to conserve fish stocks.

Producing mycoprotein

If good, protein-rich food can be produced in other ways, it might help to reduce the need for fishing. Mycoprotein is now widely available in the UK.

To make mycoprotein, the fungus *Fusarium* is grown in huge vats, containing glucose on which it feeds. The glucose is usually made from waste, left over from processing other food – such as starch from making flour.

Conditions inside the fermenter are adjusted to make sure that the fungus can grow in an ideal environment.

The contents are not stirred, because the fungus is made up of **hyphae** – long, thin threads. These would tangle and break. Instead, air is bubbled through the mixture. This not only supplies oxygen for aerobic respiration, but also helps to mix the contents.

The fungus is harvested and purified before it is prepared for sale. Its hyphae give it a texture a little like meat.

It is an excellent food, high in protein and fibre and low in fat.

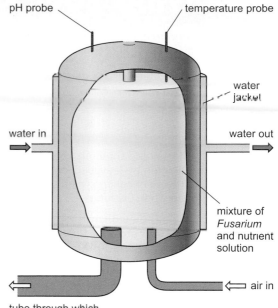

FIGURE 3: This mycoprotein fermenter has instruments to measure pH and temperature. Why is this?

QUESTIONS

4 Suggest some advantages of producing mycoprotein, rather than farming animals or catching wild fish.

5 Suggest why the fermenter in Figure 3 has a water cooler around it.

6 Eating mycoprotein might be better for a person's health than eating meat. Explain why.

Pressures on fish stocks

Scientists have worked out that there needs to be 150 000 cod and 140 000 haddock in the North Sea, for the populations to continue to survive.

The graph, Figure 4, shows what happened to the numbers of three species of fish in the North Sea between 1963 and 2010.

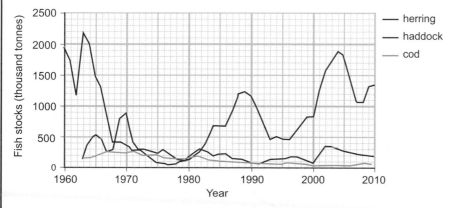

FIGURE 4: North Sea stocks of cod, herring and haddock.

QUESTIONS

7 Describe the changes in the population, since 1963, of (a) cod (b) herring.

8 Use the data in Figure 4 to suggest what may happen to the cod and haddock populations.

9 Cod is the favourite fish of many people in Britain. Suggest what can be done to reduce the pressures on North Sea cod populations.

Preparing for assessment: Applying your knowledge

To achieve a good grade in science, you not only have to know and understand scientific ideas, but you need to be able to apply them to other situations and investigations. These tasks will support you in developing these skills.

✺ On the treadmill

Zhahid is fifteen and thinking about what he is going to do when he leaves school. He goes along to the local tertiary college, with some of his friends, to an open evening.

One of the areas they go into is called a Human Performance Laboratory. It does not look much like a laboratory. It is more like the fitness suite at the local sports centre, with lots of exercise equipment and staff in tracksuits carrying clipboards.

"Right then," said one of the staff, "My name's Liz. Let's show you some of the tests that we can do on people here." She took them over to a machine that looked like a conveyor belt with handrails.

"This is a treadmill," she said, "OK, who is into running?" Nobody said anything, but they all, apart from Zhahid, took a step backward.

"Fine," said Liz, "Come on then, on you get." Zhahid stood on the treadmill and Liz attached a strap and a tube around his chest.

"This will measure your breathing rate and your skin temperature," she said. She pressed a button and the treadmill started to move. Instinctively Zhahid started to walk. Numbers appeared on the display panel. Liz turned a control and the belt started to move faster. Zhahid broke into a gentle jog.

"How does that feel?" she asked.

"Fine," he answered. She turned the control again. He was now running.

"Think about the changes that are going on in Zhahid's body now," Liz said to the others. "It is responding to this situation. Think about what is going on with the arteries and muscles."

They watched the numbers as the speed of the treadmill was increased again.

After 10 minutes, Liz said, "I'll bring it back down now."

"That was really good, Zhahid. You are obviously fit." The treadmill slowed down and Zhahid's run became a jog, then a walk and finally he stopped. She unfastened the strap and tube.

"That was alright," thought Zhahid. "I wouldn't mind doing something on that next year."

✳ Task 1

(a) What would be happening to Zhahid's breathing as he ran faster?

(b) Explain why that change was taking place.

✳ Task 2

The temperature of Zhahid's skin increased after he had been running for a few minutes.

Explain why that happened.

✳ Task 3

Liz was hoping to get a new sensor to measure rate of sweating.

(a) How would you expect Zhahid's rate of sweating to change when he ran?

(b) Why would that happen?

✳ Task 4

Zhahid had felt a little hungry before he started running, and afterwards he felt really hungry. "You have probably used up some of your glycogen stores." said Liz. "Go and eat something starchy."

Explain what Liz meant, and why eating starchy food might be a good idea.

✳ Maximise your grade

Answer includes showing that you...	
E	can identify two changes to Zhahid's breathing.
	can explain why Zhahid's skin temperature increased.
	can predict what would happen to Zhahid's rate of sweating.
	can explain why Zhahid's breathing changes took place.
C	can explain, with reference to respiration and the blood, why Zhahid's skin temperature increased.
	can explain the changes to Zhahid's rate of sweating.
A	can explain the effects of running and eating starchy food on glycogen stores, with reference to glucagon and insulin.

67

Checklist B3.3–3.4

To achieve your forecast grade in the exam you will need to revise

Use this checklist to see what you can do now. Refer back to the relevant topics in this book if you are not sure. Look across the three columns to see how you can progress. Bold text means Higher tier only.

Remember that you will need to be able to use these ideas in various ways, such as:

> interpreting pictures, diagrams and graphs

> applying ideas to new situations

> explaining ethical implications

> suggesting some benefits and risks to society

> drawing conclusions from evidence you are given.

Look at pages 188–209 for more information about exams and how you will be assessed.

To aim for a grade E	To aim for a grade C	To aim for a grade A
Recall some of the waste products that need to be removed from the human body.	Describe the source of waste products in the human body and how they are removed.	Explain in detail how the kidney filters the blood and how this contributes to homeostasis.
Recall a reason for and consequence of kidney failure.	Describe in simple terms the two treatments for kidney failure.	Explain the process of dialysis. Compare the advantages and disadvantages of dialysis with those of transplants.
Recall the normal body temperature.	Describe how the body temperature is monitored. Describe the role of sweat in cooling the body.	**Understand the role of the skin in temperature regulation.**
Know that glucose is important as a fuel for body cells, enabling them to produce energy.	Describe the role of the pancreas in controlling blood glucose.	**Explain the action of both insulin and glucagon in controlling blood glucose.**
Know that, if the blood glucose is not controlled, then a person will have a disease called diabetes.	Describe the medical and lifestyle treatments for Type 1 diabetes.	

To aim for a grade E

Recall that waste can pollute water, land and the air.

Recall that deforestation happens on a large scale in tropical regions.

Recall that deforestation increases the carbon dioxide in the atmosphere.

Know that increasing carbon dioxide and methane in the air contributes to global warming.

Understand that biofuels are made from living organisms.

Describe how energy is passed from one stage to another in a food chain.

Recognise that food does not necessarily come from farms and that it can come from the wild – for example, fishing.

To aim for a grade C

Describe specific examples of how waste can pollute water, land and the air.

Explain why deforestation takes place and how it affects animal and plant life.

Describe how deforestation can affect the atmospheric carbon dioxide level.

Describe possible effects of global warming.

Describe how a range of different biofuels are made.

Explain why a reduction in the number of stages in a food chain increases the efficiency of food production.

Explain how the fishing industry can be made into an example of sustainable food production.

To aim for a grade A

Describe the link between human population growth, pollution and land use.

Understand that deforestation in tropical regions cannot be described as sustainable use of trees.

Explain how the destruction of peat bogs contributes to increasing the atmospheric carbon dioxide level.

Describe how natural processes play a part in slowing down the increase in carbon dioxide.

Explain how biofuel use could contribute to slowing down the global rise in atmospheric carbon dioxide.

Explain how intensive animal farming techniques can improve the efficiency of food production.

Describe how protein-rich food can be produced from fungi.

1. People who suffer from kidney failure may have a kidney transplant. They usually have to take immunosuppressant drugs for the rest of their lives. This stops their immune systems from attacking and destroying the transplanted kidney.

A study, carried out in Australia, looked at the number of cases of cancer in people who had kidney transplants, compared with people who had not had transplants. The results are shown in the graph.

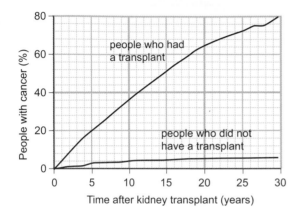

AO2 **(a)** To collect the data for the 'people who had not had a transplant' line on the graph, the researchers used people who were the same age as the people who had had a transplant.

Suggest why they did this. [2]

AO3 **(b)** Use the information in the graph to describe how having a kidney transplant affects the risk of getting cancer. [3]

AO2 **(c)** The body's immune system often destroys cancer cells before they have a chance to produce a tumour.

Suggest how this can explain your answer to part (b). [2]

AO3 **(d)** In the study, most of the cancers in the people who had had a kidney transplant were skin cancer. If detected early, skin cancer can usually be treated successfully.

Use the information above, and your own knowledge, to suggest why a person with kidney disease might prefer to have a kidney transplant rather than having kidney dialysis. [3]

2. Borneo is a large island in south-east Asia. Since the middle of the 20th century, large areas of tropical rainforest have been cleared to make way for growing oil palm trees.

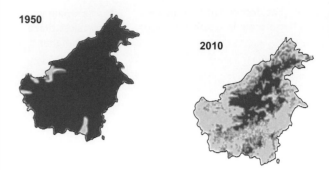

Tropical rainforest in Borneo

(a) These maps show how the area of tropical rainforest in Borneo changed between 1950 and 2010. The green areas indicate forest.

AO3 **(i)** Describe the change in the area of tropical rainforest in Borneo between 1950 and 2010. [2]

AO1 **(ii)** Explain the meaning of the term 'biodiversity'. [1]

AO2 **(iii)** Explain how the change that you have described is likely to lead to a reduction in biodiversity. [3]

(b) The palm trees are used to produce palm oil. Some of the palm oil is used to produce biodiesel, which is used as fuel.

AO2 **(i)** Suggest how the use of biofuels made from palm oil, rather than fossil fuels, may help to reduce global warming. [3]

AO2 **(ii)** Suggest how biofuels could be produced in a way that would cause less harm to the environment than growing oil palms. [2]

AO1 recall the science   AO2 apply your knowledge   AO3 evaluate and analyse the evidence

✳ WORKED EXAMPLE – Foundation tier

In this question you will be assessed on using good English, organising information clearly and using specialist terms where appropriate.

Much of our food is produced by farmers, who grow crops and keep animals from which food is made. As the human population increases, many people think it is important that we try to increase the efficiency of food production, and to limit food miles.

Explain how the efficiency of food production can be increased, and describe how limiting food miles can help to reduce damage to the environment. [6]

Eating plants is better than eating animals. This is because energy is lost as you go along a food chain.

Plants are at the start of the food chain, so not as much energy has been lost. For example, we can get more energy by eating maize grown in a field, instead of eating cattle that have eaten the maize.

Farmers can also increase efficiency by keeping their animals indoors where it is warm, and not letting them move around too much. This means the animals can use more of their food for growing, rather than using it for energy to move around or keep warm. So the farmer gets more beef or milk for each kg of food he gives his animals.

Food miles are how far your food has travelled before it gets to you. Moving food around uses up fuel, which can add to the carbon dioxide in the air and cause global warming.

How to raise your grade!
Take note of these comments – they will help you to raise your grade.

⬇

This question asks about two things. The first part of the question is about efficiency of food production; the second part is about food miles.

This is a good start. The candidate has written down useful information straight away – no words have been wasted.

These are good points about how a farmer can increase efficiency of food production from animals.

There is another good point, describing what 'efficiency' means – the quantity of useful products (in this case, beef or milk) compared with the amount of 'inputs' (in this case, food for the animals) that has been put in.

This is a brief explanation of one of the benefits of reducing food miles, and it is correct. The candidate could have stated that moving food more miles will increase the problem/increase global warming.

This candidate will be given five or six marks.

The answer has been planned carefully. Each sentence says something new, relevant and useful. There is no repetition. Spelling, punctuation and grammar are good. The answer is divided into four paragraphs, each one covering a different idea.

You should always try to use technical terms. The candidate could have included 'biomass', and perhaps 'factory farming'.

1. In London in the mid 20th century, coal was often used to heat houses. In December 1952, the weather was foggy. Sulfur dioxide, produced when the coal was burned, was trapped at ground level by the fog.

The graph shows the number of people who died in London between 1 and 15 December, 1952. It also shows the concentration (in parts per million) of sulfur dioxide in the air.

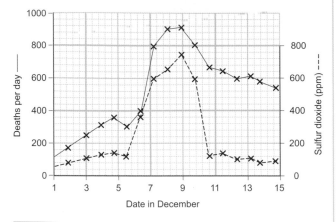

AO3 **(a)** Calculate the increase in deaths per day between 4 December and 8 December. [1]

AO3 **(b)** Discuss the extent to which the data shown in the graph support the idea that sulfur dioxide is harmful to human health. [4]

AO1 **(c)** Sulfur dioxide can harm plants and aquatic animals. Explain how this happens. [2]

2. Read the passage, below, about measures to conserve fish stocks, then answer the questions that follow.

In March 2011, the European Union announced plans to ban the discarding of unwanted fish back into the sea.

Up to now, in order to achieve sustainable food production, regulations have imposed quotas of particular fish species on fishermen. If they caught more than their quota of that species, perhaps while fishing for a different species, they were not allowed to bring the over-quota fish back to land. They had to throw the dying fish back into the sea.

Gulls follow fishing boats, as shown in the photograph, both in harbour and far out at sea, to feed on this waste.

The new proposal hopes to regulate fish stocks by limiting the time for which a fishing boat can catch fish. However, many people working in the fishing industry do not agree with the proposals, because this could greatly limit the income of fishing fleets and perhaps put them out of business. They say that they have already been working hard to reduce discards, and that using measures such as limiting the size of the mesh of fishing nets, or allowing people to fish only in certain areas, are already helping to solve the problem.

AO3 **(a)** Describe what is meant by the term 'sustainable food production'. [2]

AO3 **(b)** Explain how each of the following can help in the conservation of fish stocks.

(i) Imposing fishing quotas. [2]

(ii) Regulating the mesh size of fishing nets. [2]

AO2 **(c)** Use the information in the passage, and your own knowledge, to suggest why fishermen have had to discard some of their catch. [3]

AO2 **(d)** Suggest how banning fishermen from discarding fish could help to conserve stocks of fish species that are under threat. [2]

AO2 **(e)** Fish are an excellent source of protein in the diet. One way of helping to conserve fish stocks might be to persuade people to eat more mycoprotein instead of fish.

Suggest reasons why the manufacture of mycoprotein can be considered to be an example of sustainable food production. [2]

AO1 recall the science   AO2 apply your knowledge   AO3 evaluate and analyse the evidence

✳ WORKED EXAMPLE – Higher tier

In this question you will be assessed on using good English, organising information clearly and using specialist terms where appropriate.

The concentration of glucose in a person's blood is normally kept fairly constant. In Type 1 diabetes, this control system does not work correctly.

Describe how blood glucose concentration is normally controlled, and explain the measures that a person with Type 1 diabetes should take, to control their blood glucose concentration. [6]

Blood glucose is controlled by insulin and glycogon.

When you have too much glucose in the blood, insulin makes it go down. When you haven't got enough, glycogon makes it go up. The pancreas and liver help with this.

If you have diabetes, then you don't have enough insulin and you may need to inject it. If you don't do that, your blood glucose may go very high and you may go into a coma. In between meals, your blood glucose can go very low and this makes you feel bad-tempered and perhaps faint. So you have to eat sugary things if this happens.

If you do lots of exercise, you need to make sure you eat more sugar.

How to raise your grade!
Take note of these comments – they will help you to raise your grade.

Notice that this question has two parts: 'Describe' and 'Explain'.

There is no such substance as 'glycogon'. The candidate has muddled the hormone glucagon with the storage substance glycogen.

It is very important to use and spell key terms correctly.

Examiners do not penalise twice for the same mistake, so the wrong spelling here will not affect the mark.

The candidate has mentioned the pancreas and the liver but much more explanation is needed.

The candidate could have said that the pancreas does not make enough insulin, rather than just saying that there is not enough.

There is a very brief outline of what happens if blood glucose concentrations rise too high or fall too low. The candidate could have said more about why this happens.

The candidate has mentioned exercise in the answer. However, it is not a very full explanation.

Overall, this is a fairly average answer and receives three marks out of six.

The candidate has tried to answer all the parts of the question. They have structured it well. Apart from the wrong spelling of 'glucagon', spelling and grammar are good. However, the descriptions and explanations are very shallow. A good answer would give a lot more detail.

Chemistry C3.1–3.3

What you should know

The periodic table

The periodic table arranges elements in groups and periods, according to their properties.

The elements in a group have the same number of electrons in their highest energy level and have similar properties.

- Name a group of elements in the periodic table and describe one way in which their properties are similar.

Salts, solubility, limestone and ions

Salts contain ions.

Salts may be soluble or insoluble in water.

Insoluble salts form as precipitates in a reaction.

Soluble salts form a mixture of their ions in a solution.

- What is the chemical name and formula for limestone?

Energy changes in a chemical reaction

Chemical reactions are usually accompanied by an energy change. This may be exothermic or endothermic.

Exothermic reactions transfer energy to the surroundings.

Endothermic reactions take in energy from the surroundings.

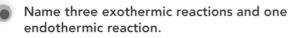

- Name three exothermic reactions and one endothermic reaction.

The environment

Carbon dioxide is one of the greenhouse gases responsible for global warming.

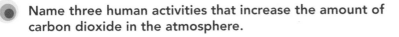

- Name three human activities that increase the amount of carbon dioxide in the atmosphere.

You will find out

The periodic table

> Newlands and Mendeleev devised early periodic tables.

> Mendeleev's periodic table was the forerunner of the modern periodic table.

> The modern periodic table has the elements arranged in terms of their electronic structures.

Water

> The substances dissolved in tap water determine whether the water is hard or soft.

> Different methods are used to soften temporary and permanent hard water.

> Water filters may be used to improve the quality of tap water.

> Chlorine and fluorine may be added to water supplies.

> Distillation can be used to produce pure water.

Calculating and explaining energy changes

> Calorimetry is used to calculate the energy given out when a food or fuel combusts.

> Energy level diagrams follow the energy changes in a reaction.

> Bond energies can be used to calculate the energy change in a reaction.

> Hydrogen may be the fuel of the future.

Ordering elements

You will find out:

> how John Newlands sorted the elements into an early periodic table

> why Mendeleev is known as the father of the periodic table

> how early periodic tables became a useful tool

Tunes and elements

Online stores organise music into groups such as pop, rock and classical. This makes it easier to find the music that you like and predict its style. In the 19th century, discovering a new element was a regular event. Scientists started looking for patterns in the elements' properties. They needed a way to sort elements into groups.

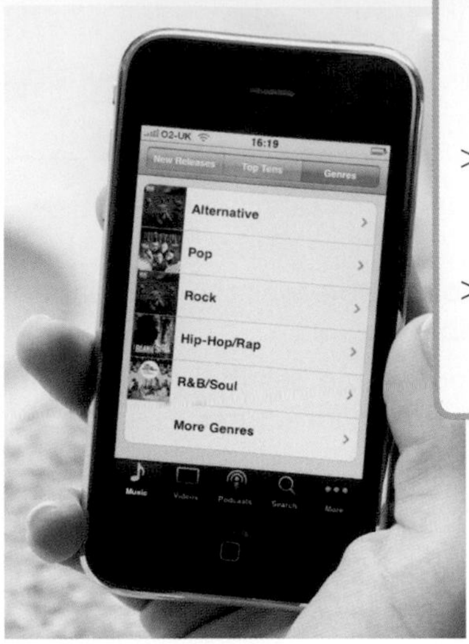

FIGURE 1: Categories of music make searching easier.

Ordering and sorting

Newlands

John Newlands worked as a chemist in a sugar factory. By 1864, about 60 elements had been discovered. Newlands used atomic weights (relative atomic masses) to sort these known elements into a table. (Scientists at the time did not know about protons, electrons and neutrons.) He realised that elements with similar properties appeared every eight places. Therefore, he arranged his table in columns of seven elements. This gave eight columns. He called it **The Table of Octaves**.

H	1	F	8	Cl	15	Co/Ni	22	Br	29	Pd	36	I	42	Pt/Ir	50
Li	2	Na	9	K	16	Cu	23	Rb	30	Ag	37	Cs	44	Tl	53
Gl	3	Mg	10	Ca	17	Zn	25	Sr	31	Cd	34	Ba/V	45	Pb	54
B	4	Al	11	Cr	18	Y	24	Ce/La	33	U	40	Ta	46	Th	56
C	5	Si	12	Ti	19	In	26	Zr	32	Sn	39	W	47	Hg	52
N	6	P	13	Mn	20	As	27	Di/Mo	34	Sb	41	Nb	48	Bi	55
O	7	S	14	Fe	21	Se	28	Ro/Ru	35	Te	43	Au	49	Os	51

TABLE 1: Newland's listed the elements vertically in his *Table of Octaves*. Different symbols are used now for some elements. What elements are Gl and Bo?

John Newlands had a few problems with his Table of Octaves.

> Many of the atomic weights he used were not accurate.

> Some of his 'elements' were actually mixtures of two elements.

> He did not make room for elements not yet discovered.

> He put two elements in some boxes when convenient.

> He assumed the pattern must repeat itself every eight elements, so iron ended up in the same group as oxygen and sulfur.

Mendeleev

In 1869, Dmitri Mendeleev, a Russian lecturer, published his periodic table. He used cards to represent each element as a teaching aid – and discovered that the elements could be arranged in a pattern that linked their properties to their masses.

QUESTIONS

1 Which group of elements is missing from both Newland's Table of Octaves and Mendeleev's periodic table?

2 Compare Newland's *Table of Octaves* with Mendeleev's periodic table.

Describe how the tables are (a) similar (b) different.

... law of octaves ... John Newlands

> Mendeleev put the elements in the correct order. He realised that some atomic weights must be wrong and recalculated them.

> He lined up the elements with similar properties. These are **groups**. Mendeleev placed tellurium before iodine, even though iodine has a higher atomic weight than tellurium. Iodine's properties are very similar to the properties of fluorine and chlorine. Tellurium has properties similar to oxygen and sulfur.

> Mendeleev realised that the repeating pattern was not always after eight elements.

> He left gaps for the elements yet to be discovered.

			K = 39	Rb = 85	Cs = 133	—	—
			Ca = 40	Sr = 87	Ba = 137	—	—
			—	?Yt = 88?	?Di = 138?	Er = 178?	—
			Ti = 48?	Zr = 90	Ce = 140?	?La = 180?	Tb = 231
			V = 51	Nb = 94	—	Ta = 182	—
			Cr = 52	Mo = 96	—	W = 184	U = 240
			Mn = 55	—	—	—	—
Typische Elemente			Fe = 56	Ru = 104	—	Os = 195?	—
			Co = 59	Rh = 104	—	Ir = 197	—
			Ni = 59	Pd = 106	—	Pt = 198?	—
H = 1	Li = 7	Na = 23	Cu = 63	Ag = 108	—	Au = 199?	—
	Be = 9,4	Mg = 24	Zn = 65	Cd = 112	—	Hg = 200	—
	B = 11	Al = 27,3	—	In = 113	—	Tl = 204	—
	C = 12	Si = 28	—	Sn = 118	—	Pb = 207	—
	N = 14	P = 31	As = 75	Sb = 122	—	Bi = 208	—
	O = 16	S = 32	Se = 78	Te = 125?	—	—	—
	F = 19	Cl = 35,5	Br = 80	J = 127	—	—	—

FIGURE 2: Mendeleev listed the elements vertically. The groups go across the table. Notice the gaps. Which element goes in the gap next to silicon?

His table was called a periodic table because similar properties occur at regular intervals, or periods. Mendeleev's table was the forerunner of the modern periodic table in use today.

Breaking the code

Mendeleev not only left gaps in his table for undiscovered elements, he predicted the properties of the missing elements.

Mendeleev called the yet-to-be-discovered element below aluminium 'eka-aluminium'. Eka-aluminium was discovered in 1875 and called gallium. Table 2 shows Mendeleev's predicted properties and the real properties for gallium.

Property	Predicted	Known
atomic weight	about 68	69.72
density (g/cm³)	6.0	5.9
melting point (°C)	low	29.78
formula of oxide	Ea_2O_3	Ga_2O_3
oxide solubility	in acids and alkalis	in acids and alkalis

TABLE 2: Properties of gallium – comparing Mendeleev's prediction with its known properties today

Three of Mendeleev's predicted elements were discovered in his lifetime. Each time, his predictions proved to be accurate. Mendeleev received one of the greatest honours – to have an element named after him. Element 101 is Mendelevium.

FIGURE 3: The metal is gallium. Which of the properties predicted by Mendeleev does this demonstrate?

QUESTIONS

3 What is the modern name for eka-boron?

4 What did Mendeleev call germanium in his periodic table?

The noble gases

Noble gases undergo just a handful of chemical reactions. The highest electron energy level of their atoms is full. They do not need to lose or gain electrons to have a stable electronic configuration.

Their unreactivity made them difficult to discover. Helium was isolated in 1868 and, by 1898, the first five noble gases had been discovered. Mendeleev included them in his periodic table by adding an extra row for the noble gases. This is Group 0.

QUESTIONS

5 Explain why Mendeleev's periodic table has become a useful tool.

The modern periodic table

The race goes on

Element number 117 was discovered in 2010 and given the name ununseptium. It was made by scientists in a particle accelerator and lasted for milliseconds before it decayed. Before then, it had never existed on Earth and probably never in the history of the Universe.

Did you know?

When Mendeleev discovered his periodic table, new elements were being discovered by using chemistry. Today, new elements are made in particle accelerators.

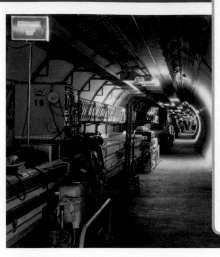

FIGURE 1: The Large Hadron Collider at Cern, Switzerland is a particle accelerator.

You will find out:

> elements in the modern periodic table are arranged in order of atomic number
> an element's position in the periodic table shows its electron configuration
> elements in the same group have the same number of electrons in their highest energy level (shell)

The modern periodic table

Electrons, protons and neutrons were discovered in the early 20th century.

The modern periodic table arranges the elements in order of increasing **atomic number** rather than relative atomic mass.

> Hydrogen, with one proton in its atom, is the first element in the table.

> Helium with two protons is the second, and so on.

> The **groups** are numbered from 1 to 7, with the noble gases in Group 0.

> Elements across a **period** show a gradual change from metal to non-metal. For example, in period 3:

> sodium is a typical metal
> chlorine is a typical non-metal, but
> silicon has both some non-metal and some metal properties.

QUESTIONS

1 Does the relative atomic mass of elements always increase as you go through the modern periodic table?

2 Compare Mendeleev's table with the modern periodic table.

Describe how the tables are (a) similar (b) different.

Group 1	2											Group 3	4	5	6	7	0
		$^{1}_{1}$H hydrogen															$^{4}_{2}$He helium
$^{7}_{3}$Li lithium	$^{9}_{4}$Be beryllium											$^{11}_{5}$B boron	$^{12}_{6}$C carbon	$^{14}_{7}$N nitrogen	$^{16}_{8}$O oxygen	$^{19}_{9}$F flourine	$^{20}_{10}$Ne neon
$^{23}_{11}$Na sodium	$^{24}_{12}$Mg magnesium											$^{27}_{13}$Al aluminium	$^{28}_{14}$Si silicon	$^{31}_{15}$P phosphorus	$^{32}_{16}$S sulfur	$^{35}_{17}$Cl chlorine	$^{40}_{18}$Ar argon
$^{39}_{19}$K potassium	$^{40}_{20}$Ca calcium	$^{45}_{21}$Sc scandium	$^{48}_{22}$Ti titanium	$^{51}_{23}$V vanadium	$^{52}_{24}$Cr chromium	$^{55}_{25}$Mn manganese	$^{56}_{26}$Fe iron	$^{59}_{27}$Co cobalt	$^{59}_{28}$Ni nickel	$^{64}_{29}$Cu copper	$^{65}_{30}$Zn zinc	$^{70}_{31}$Ga gallium	$^{73}_{32}$Ge germanium	$^{75}_{33}$As arsenic	$^{79}_{34}$Se selenium	$^{80}_{35}$Br bromine	$^{84}_{36}$Kr krypton
$^{85}_{37}$Rb rubidium	$^{88}_{38}$Sr strontium	$^{89}_{39}$Y yttrium	$^{91}_{40}$Zr zirconium	$^{93}_{41}$Nb niobium	$^{96}_{42}$Mo molybdenum	$^{99}_{43}$Tc technetium	$^{101}_{44}$Ru ruthenium	$^{103}_{45}$Rh rhodium	$^{106}_{46}$Pd palladium	$^{108}_{47}$Ag silver	$^{112}_{48}$Cd cadmium	$^{115}_{49}$In indium	$^{119}_{50}$Sn tin	$^{122}_{51}$Sb antimony	$^{128}_{52}$Te tellurium	$^{127}_{53}$I iodine	$^{131}_{54}$Xe xenon
$^{133}_{55}$Cs caesium	$^{137}_{56}$Ba barium	$^{139}_{57}$La lanthanum	$^{178}_{72}$Hf hafnium	$^{181}_{73}$Ta tantalum	$^{184}_{74}$W tungsten	$^{186}_{75}$Re rhenium	$^{190}_{76}$Os osmium	$^{192}_{77}$Ir iridium	$^{195}_{78}$Pt platinum	$^{197}_{79}$Au gold	$^{201}_{80}$Hg mercury	$^{204}_{81}$Tl thallium	$^{207}_{82}$Pb lead	$^{209}_{83}$Bi bismuth	$^{210}_{84}$Po polonium	$^{210}_{85}$At astatine	$^{222}_{86}$Rn radon
$^{223}_{87}$Fr francium	$^{226}_{88}$Ra radium	$^{227}_{89}$Ac actinium															

FIGURE 2: The modern periodic table is a useful tool for organising the building blocks of the Universe.

FIGURE 3: A box in a modern periodic table.

… modern periodic table GCSE

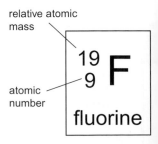

Electron configurations

Configurations and the periodic table

The number of protons in an atom is the same as the number of electrons (remember that atoms are electrically neutral) – the atomic number also equals the number of electrons.

> Hydrogen has one electron in the first energy level.

> Helium, in period 1, has two electrons.

> Lithium has three electrons – two in the first energy level and one in the second energy level. Across period 2, one electron is added for each element. Neon has eight electrons in the second energy level. This is the maximum number of electrons in this energy level.

> For elements in period 3, electrons fill the third energy level in a similar way.

Configurations and groups

Electrons in the outer (highest) energy level are rearranged when chemical reactions happen. The number of electrons in the highest energy level is linked to an element's chemical properties.

All the elements in a group have the same number of electrons in their highest energy level. The number of electrons is also the group number.

For example, fluorine and chlorine are both in Group 7 and have 7 electrons in their highest energy levels. When they react, each atom needs to gain one electron to become stable. Fluorine and chlorine have similar chemical properties, as do all the elements in Group 7.

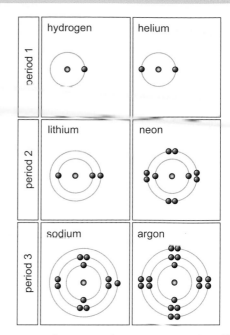

FIGURE 4: Electron configurations. Note how filling a new energy level starts a new period in the periodic table. What other patterns can you see?

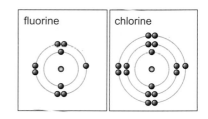

FIGURE 5: Electron configurations of Group 7 fluorine and chlorine. How are fluorine and chlorine atoms similar and how are they different?

⦿ QUESTIONS

3 Group 2 elements have how many electrons in their highest energy level?

4 Draw the electron configurations for (a) boron (b) oxygen (c) magnesium.

A work in progress

Elements up to uranium (atomic number 92) in the periodic table occur naturally. Heavier elements are usually radioactive and decayed away long ago, if they ever existed on Earth. Scientists make the heavier elements in particle accelerators.

Neptunium (atomic number 93) was the first one made in an accelerator, by bombarding uranium atoms with neutrons. When scientists around the world agree that a heavy element has been successfully made, it is added to the periodic table.

Three atoms of ununoctium (atomic number 118) were made in 2006. They lasted for less than a millisecond. Ununseptium was made in 2010. Scientists are now trying to make even heavier elements.

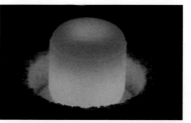

FIGURE 6: Plutonium is a really useful element, used in nuclear reactors. Where do supplies of plutonium come from?

⦿ QUESTIONS

5 If scientists discover more stable, super heavy elements, which last for a few hours, suggest what difference it will make to science.

Group 1

You will find out:
> trends in physical properties of Group 1 metals
> trends in reactivity of Group 1 metals
> how their electron configuration explains their reactivity trends

Curious metals

Imagine a group of metals so light that many float on water, so soft that many can be cut with a knife, and so reactive that they must be stored under oil. These are not the properties of ordinary metals. These are the Group 1 metals, lithium, sodium, potassium, rubidium and caesium. They are the alkali metals.

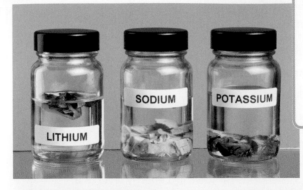

FIGURE 1: Oil prevents water and oxygen reacting with these metals.

Did you know?

Hydrogen is a special case and usually considered on its own. Like all Group 1 elements, hydrogen's atoms have one electron in their highest energy levels and, like all Group 7 elements, hydrogen atoms need one electron to fill up their highest energy levels.

What are they like?

Physical properties

Group 1 elements are silvery shiny metals when freshly cut.

The density of lithium, sodium and potassium is less than the density of water (1.00 g/cm³): they float on water.

Element	Atomic number	Relative atomic mass	Melting point (°C)	Density (g/cm³)
Li	3	7	181	0.53
Na	11	23	98	0.97
K	19	39	63	0.86
Rb	37	85	39	1.53
Cs	55	133	29	1.88

TABLE 1: Some properties of Group 1 elements.

FIGURE 2: The freshly cut surface of sodium.

Chemical reactions

> Atoms of Group 1 elements all have one electron occupying their highest energy level. When they react, they transfer this electron to leave an ion with one positive charge, +1. For example:

Na → Na⁺ + e⁻
(Na⁺ is a sodium ion and e⁻ is the transferred electron.)

> Group 7 elements, such as chlorine, have seven electrons in their highest energy level. When sodium reacts with chlorine, one electron is transferred from a sodium atom to a chlorine atom and an ionic bond forms. A sodium ion and a chloride ion (Cl⁻) are made.

$2Na(s) + Cl_2(g) \rightarrow 2NaCl(s)$

FIGURE 3: Potassium reacting with water. What would you expect to see if universal indicator was added to the water?

Q ... group 1 metals ... group 1 alkali metals periodic table

Lithium and potassium also react with Group 7 elements to form halides (an iodide, chloride or bromide). The pattern is the same: A white crystalline salt forms that is soluble in water.

> Group 1 metals all react with cold water. In each case, hydrogen gas is given off and a metal hydroxide made. The hydroxide dissolves in the water to make an alkaline solution. This is why Group 1 elements are also called alkali metals.

As you go down Group 1, the metals react increasingly violently with water. Caesium explodes in water and can shatter the container.

Group 1 metal	Observations	Equation
lithium	moves around, fizzes	$2Li(s) + 2H_2O(\ell) \rightarrow 2LiOH(aq) + H_2(g)$
sodium	moves around more violently, fizzes, may spit	$2Na(s) + 2H_2O(\ell) \rightarrow 2NaOH(aq) + H_2(g)$
potassium	moves around violently hydrogen given off burns with a lilac flame	$2K(s) + 2H_2O(\ell) \rightarrow 2KOH(aq) + H_2(g)$

TABLE 2: Examples of how Group 1 elements react with water.

Remember

The higher the energy level of the outer electrons, the more easily they are lost and the more reactive the metal.

QUESTIONS

1 Describe the trend in melting points and density as you go down Group 1.

2 State which Group 1 metals would be liquid at 100 °C.

3 Write word and symbol equations for the reaction of caesium with water.

Explaining the trends (Higher tier)

Group 1 metals become more reactive as you go down the group:

> They all have one electron in their highest energy level, but each successive metal has an extra inner energy level filled with electrons.

> The atoms are larger.

> Negatively charged electrons are attracted to the positively charged nucleus. This keeps the electrons in place.

> The smaller atoms have fewer filled energy levels between the outer electron and the nucleus. The electron is not easily lost.

> The larger atoms have more filled energy levels to shield the attraction of the positively charged nucleus and so the negatively charged electrons are lost more easily.

electron in highest energy level is more easily lost →

number of energy levels filled with electrons increases →

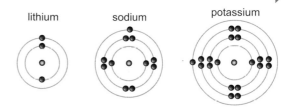

lithium   sodium   potassium

FIGURE 4: Electron configurations of the alkali metals.

QUESTIONS

4 Explain why caesium explodes in water.

Francium

Francium is the most reactive Group 1 element. Its atomic number is 87 and its most stable isotope is francium-223.

This highly radioactive element is the second rarest naturally occurring element. Francium is found in trace amounts in uranium and thorium minerals. It has no commercial applications.

QUESTIONS

5 (i) Describe the atomic structure of a francium-223 atom.

(ii) Predict values for the melting point and density of francium.

Q ... physical properties alkali metals GCSE

Transition metals

You will find out:

> physical properties of transition metals
> how their reactivity compares with Group 1 metals
> about some properties of transition metal compounds

Paint colours

Transition metal compounds are the coloured pigments used in many paints. Cobalt and manganese compounds are used for blue and violet colours. Iron compounds give yellow, red and brown pigments.

Did you know?

Transition metals can have up to 32 electrons in one energy level.

FIGURE 1: Unlike alkali metal compounds, most transition metal compounds are brightly coloured.

Properties

Hard, tough and strong

Compared with Group 1 metals, transition metals are hard, tough and strong.

The strong metallic bonds in transition metals hold the ions tightly together and account for their hardness, strength and high densities.

High temperatures are needed to break these bonds, so transition metals have high melting and boiling points. This makes them very useful for building bridges, vehicles and machinery.

Table 1: Physical properties of some transition metals.

Metal	Melting point (°C)	Boiling point (°C)	Density (g/cm³)
titanium	1668	3287	4.54
chromium	1857	2672	7.19
manganese	1246	1962	7.33
iron	1538	2861	7.87
copper	1083	2567	8.92

FIGURE 3: The bumpers on the Cadillac are made from the transition metal chromium. Chromium's unreactivity means the bumpers do not corrode and stay shiny.

Reactivity

Transition metals are much less reactive than Group 1 metals. Iron, for example, reacts very slowly with oxygen and water to form rust. However, heated iron wool (with a high surface area) reacts with steam.

$3Fe(s) + 4H_2O(\ell) \rightleftharpoons Fe_3O_4(s) + 4H_2(g)$ This is a reversible reaction.

$^{48}_{22}$Ti	$^{51}_{23}$V	$^{52}_{24}$Cr	$^{55}_{25}$Mn	$^{56}_{26}$Fe	$^{59}_{27}$Co	$^{59}_{28}$Ni	$^{64}_{29}$Cu
titanium	vanadium	chromium	manganese	iron	cobalt	nickel	copper
$^{91}_{40}$Zr	$^{93}_{41}$Nb	$^{96}_{42}$Mo	$^{99}_{43}$Tc	$^{101}_{44}$Ru	$^{103}_{45}$Rh	$^{106}_{46}$Pd	$^{108}_{47}$Ag
zirconium	niobium	molybdenum	technetium	ruthenium	rhodium	palladium	silver
$^{178}_{72}$Hf	$^{181}_{73}$Ta	$^{184}_{74}$W	$^{186}_{75}$Re	$^{190}_{76}$Os	$^{192}_{77}$Ir	$^{195}_{78}$Pt	$^{197}_{79}$Au
hafnium	tantalum	tungsten	rhenium	osmium	iridium	platinum	gold

FIGURE 2: The transition metals in the periodic table.

QUESTIONS

1 Explain why transition metals are more useful than Group 1 metals.

2 Look at Table 1. Suggest a transition metal that might be suitable for making aeroplane engine parts. Give your reasons.

Q ... transition metals overview

Electron configuration and properties

Configuration

Transition metals are sandwiched between Groups 2 and 3.

Calcium has the electron configuration 2.8.8.2. The next element, scandium, has the electron arrangement 2.8.9.2.

The period 4 transition metals are titanium to copper. Titanium has the electron arrangement 2.8.10.2. Copper is the last transition metal, with the electron arrangement 2.8.17.2.

When transition metals react, they use electrons in the highest energy level and one or more from the next to highest energy level. A transition metal can form positive ions with different size charges. For example, iron atoms can form Fe^{2+} and Fe^{3+} ions.

Colour

Most transition metals are silvery and shiny, but their compounds and solutions of their compounds are usually coloured.

Catalysis

Many transition metals are used as catalysts to increase the rate of a chemical reaction.

> Iron catalyses the reaction of nitrogen and hydrogen (Haber process for making ammonia).

> Platinum and rhodium are used for catalytic converters in car exhaust systems.

> Nickel catalyses the hydrogenation of unsaturated fats to make margarine and other spreads.

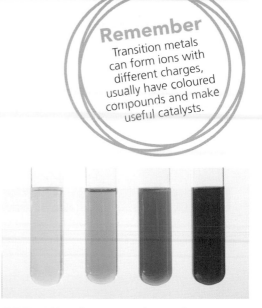

FIGURE 4: Vanadium ions with a +5 charge are yellow, +4 blue, +3 green and +2 purple.

FIGURE 5: Why does a catalytic converter need a large surface area of platinum and rhodium?

> **Remember**
> Transition metals can form ions with different charges, usually have coloured compounds and make useful catalysts.

QUESTIONS

3 Name three properties of transition metals that are different from those of alkali metals.

4 Give the electron configuration for germanium.

Highly desirable and rare

Tantalum, atomic number 73, is a dense and hard transition metal that is unreactive below 150 °C and an excellent conductor of heat and electricity. Its major use is to make capacitors for mobile phones, computers and game consoles.

Capacitors containing tantalum are very small and highly efficient. The problem is that tantalum is rare. Earth's crust contains 1.7 parts per million by mass and each mobile phone contains about 40 mg of tantalum. The demand for tantalum outstrips the supply.

FIGURE 6: Games consoles and computers rely on tantalum for small electrical components.

QUESTIONS

5 Explain why it is important to recycle mobile phones.

Group 7

Good and bad

Halogens are dangerous. Chlorine gas has been used as a chemical weapon. Fluorine is so hazardous, you will not see it in school. Yet, halogen compounds are essential to life. Iodide salts are needed to make the drug thyroxin, fluoride salts strengthen tooth enamel and chloride salts preserve food and disinfect water supplies.

You will find out:

> about the trends in the physical properties of the halogens
> how halogens react with metals and halide solutions
> why reactivity decreases down Group 7

Did you know?

Chlorine is probably the greatest life saver because of its ability to kill microbes. In most parts of the world, it has wiped out most of the diseases carried by dirty water. 160 years ago, cholera killed thousands of people every year in London. Adding chlorine to water supplies cut these deaths to zero.

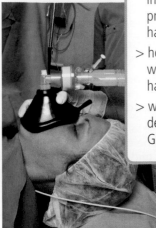

FIGURE 1: Many anaesthetics, such as halothane, are compounds made from halogens.

Properties of the halogens

Physical properties

The halogens are the colourful Group 7 non-metals. They exist as covalently bonded diatomic molecules such as Br_2 and I_2. Their properties are shown in Table 1.

Remember
The higher the energy level of the outer electrons, the less easily electrons are gained and the less reactive the halogen.

Halogen	Symbol	Atomic number	Appearance at room temperature	Melting point (°C)	Boiling point (°C)
fluorine	F	9	pale yellow gas	−219	−188
chlorine	Cl	17	yellow–green gas	−101	−34
bromine	Br	35	dark red liquid	−7	59
iodine	I	53	shiny grey solid	114	184

TABLE 1: Properties of Group 7 non-metals.

Reactions with metals

Halogens react with metals to make salts called **halides**. Chlorine makes **chlorides**, bromine makes **bromides** and iodine makes **iodides**. When sodium burns in chlorine, sodium chloride forms.

$$2Na(s) + Cl_2(g) \rightarrow 2NaCl(s)$$

Each sodium atom transfers an electron to a chlorine atom, forming a sodium ion, Na^+, and a chloride ion, Cl^-. These are held together in a giant lattice by ionic bonds.

The reaction becomes less vigorous as you go down Group 7, showing the decreasing reactivity trend.

When iron reacts with a halogen, the iron(III) halide forms. Again the reactivity of iron with Group 7 elements decreases from fluorine (most reactive) down to iodine (least reactive). For example, iron and chlorine react to form iron(III) chloride:

$$2Fe(s) + 3Cl_2(g) \rightarrow 2FeCl_3(s)$$

FIGURE 2: Chlorine, bromine and iodine at room temperature. Why is bromine seen as a liquid and a gas?

Displacement reactions

If a solution of chlorine is added to potassium bromide solution, a red–brown colour appears as bromine forms.

$$Cl_2(aq) + 2KBr(aq) \rightarrow 2KCl(aq) + Br_2(aq)$$

The ionic equation shows the essential chemistry.

$$Cl_2(aq) + 2Br^-(aq) \rightarrow 2Cl^-(aq) + Br_2(aq)$$

The chlorine has displaced the bromine from its salt. This is a **displacement reaction**. Displacement reactions also show the order of reactivity of the halogens. The reactivity decreases as you go down Group 7.

> ### QUESTIONS
>
> **1** Describe the trend in melting and boiling point as you go down Group 7.
>
> **2** Name the halogens that would be solids at (a) 100 °C (b) 100 °C?
>
> **3** Write a symbol and word equation for the reaction between bromine water, $Br_2(aq)$, and potassium iodide solution, $KI(aq)$.

FIGURE 3: What forms when aluminium reacts with iodine?

Explaining the trends (Higher tier)

All Group 7 atoms have seven electrons in their highest energy level. They gain one electron to become a halide ion, with a noble gas electron arrangement. This extra electron is attracted by the positively charged halogen atom's nucleus.

The larger Group 7 atoms have more energy levels filled with electrons. They shield the attraction better and so do not gain electrons as easily. This means that they are less reactive.

> ### QUESTIONS
>
> **4** (i) Describe how the halogens differ from the alkali metals in their pattern of reactivity.
>
> (ii) Explain your answer using electron configurations.

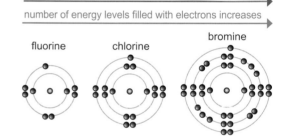

ability to gain an extra electron decreases →

number of energy levels filled with electrons increases →

fluorine   chlorine   bromine

FIGURE 4: Halogen reactivity.

Same formula, different bonding

Hydrogen gas burns in chlorine gas, reacting to form hydrogen chloride gas.

Hydrogen chloride gas is covalently bonded. It has a pungent smell and quickly becomes highly corrosive if inhaled. The covalently bonded gas dissolves in the wet lining of your nose and respiratory system and ionises into hydrogen ions and chloride ions. Hydrochloric acid is made, which corrodes the lining of your nose.

Making hydrogen chloride gas: $H_2(g) + Cl_2(g) \rightarrow 2HCl(g)$

Making hydrochloric acid: $HCl(g) + aq \rightarrow H^+(aq) + Cl^-(aq)$

The hydrogen ions cause the acidic properties.

> ### QUESTIONS
>
> **5** Describe, with the aid of diagrams, how the electrons rearrange when hydrogen chloride gas is added to water.

Hard and soft water

You will find out:
> how hard and soft water are different from each other
> about temporary and permanent hard waters
> how hard water may be softened
> about the advantages and disadvantages of hard water

A bottled choice

Some people prefer bottled water to tap water. Most bottled water is sourced from springs and has no added chlorine or fluorine. The water has filtered through layers of rock before reaching the surface again. It contains salts that dissolved naturally from the rocks and which, suppliers would claim, are good for your wellbeing.

FIGURE 1: What is dissolved in this bottled water?

Hard or soft?

You can tell whether your water is hard or soft by how easily it forms lather when you use soap. Soft water forms lather readily, but hard water does not. Also, hard water forms a scum with soap while soft water does not. Scum is an insoluble solid formed when soap reacts with compounds that make water hard.

Dissolved salts make water hard

Rainwater dissolves a number of salts from rocks, including calcium sulfate and magnesium sulfate.

> Limestone and similar rocks are insoluble in pure water. However, rainwater is slightly acidic because it contains dissolved carbon dioxide. Acidic rainwater reacts with calcium carbonate to produce a solution of calcium hydrogencarbonate.

$$CaCO_3(s) + CO_2(g) + H_2O(\ell) \rightarrow Ca(HCO_3)_2(aq)$$

> Calcium and magnesium salts dissolved in water make the water hard. Removing the salts softens the water.

Types of hardness

There are two types of hard water.

1. **Temporary hard water** – containing dissolved calcium hydrogencarbonate.

2. **Permanent hard water** – containing dissolved calcium sulfate and magnesium sulfate.

Both types form a scum with soap.

Modern soapless detergents form compounds with hard water, but these are soluble so no unpleasant scum forms. You still need more detergent to form lather, however. This is why most washing powders are detergents rather than soaps.

Health and hard water

The calcium compounds in hard water have some health benefits. They help build strong teeth and bones and also help to reduce heart disease.

FIGURE 2: A stream running over limestone rocks. Why is the water hard?

QUESTIONS

1 What are the disadvantages of hard water?

2 Explain why your tap water will be hard if it comes from a limestone area.

3 Give two health benefits of drinking hard water.

Q ... hard water GCSE

Water softening

Permanent and temporary hard waters

Soluble salts dissolve in water to form ions. Hardness in water is caused by calcium ions, Ca^{2+}, and magnesium ions, Mg^{2+}.

If **washing soda** (sodium carbonate) is added to hard water, it reacts with calcium and magnesium ions present to form precipitates of calcium carbonate and magnesium carbonate.

Passing hard water through **an ion exchange column** removes calcium ions and magnesium ions in water and replaces them with hydrogen ions or sodium ions.

Temporary hardness

Boiling temporary hard water softens it because dissolved calcium hydrogencarbonate decomposes, giving a precipitate of calcium carbonate. The calcium carbonate is deposited as limescale in water pipes, boilers, kettles and washing machines.

$$Ca(HCO_3)_2(aq) \rightarrow CaCO_3(s) + H_2O(\ell) + CO_2(g)$$

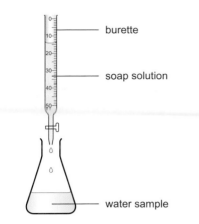

FIGURE 3: The hardness in water samples can be compared by titrating known volumes of the water sample with soap solution until a lather is obtained. Samples with higher calcium or magnesium ion concentrations will need more soap solution to produce a lather.

QUESTIONS

4 Suggest the advantages of having a water softener plumbed into your home, if you have hard water.

5 Dishwasher salt (sodium chloride) is used to recharge the ion exchange resin in a dishwasher. Explain how it works if you live in a hard water area.

Did you know?

Soap has been made from animal fat for thousands of years, but modern soapless detergents were developed only during the 1940s. Soapless detergents are less wasteful and more efficient than soap. They are used in washing powders, shampoo and shower gel, for example.

FIGURE 4: Limescale makes boilers, kettles and washing machines inefficient, so they use more electricity. Eventually, the build-up of limescale causes appliances to break down.

Ionic equations (Higher tier)

Both types of hardness form a scum with a soap (sodium stearate), $C_{17}H_{35}COONa$. For example, with calcium ions:

$$CaSO_4(aq) + 2C_{17}H_{35}COONa(aq) \rightarrow (C_{17}H_{35}COO)_2Ca(s) + Na_2SO_4(aq)$$

This can be written as an ionic equation.

$$Ca^{2+}(aq) + 2C_{17}H_{35}COO^-(aq) \rightarrow (C_{17}H_{35}COO)_2Ca(s)$$

Addition of sodium carbonate to hard water precipitates calcium carbonate and magnesium carbonate. For example, with calcium ions:

$$Na_2CO_3(aq) + CaSO_4(aq) \rightarrow CaCO_3(s) + Na_2SO_4(aq)$$

This can be written as an ionic equation.

$$CO_3^{2-}(aq) + Ca^{2+}(aq) \rightarrow CaCO_3(s)$$

QUESTIONS

6 Write ionic equations for the reactions of magnesium ions with (a) sodium carbonate solution (b) sodium stearate solution.

Q ... ionic equations

Safe drinking water

You will find out:
> what compounds may be in tap water
> advantages and disadvantages of adding chlorine and fluorine to water supplies
> how desalination and domestic water filters work

Unsafe water

In the UK, we expect safe drinking water at the turn of a tap. Yet, across the world, 4000 children die every day from diseases caused by unsafe water and sanitation. They die mainly from diarrhoeal diseases such as cholera and dysentery. The World Health Organization aims to halve the number of people with unsafe water supplies by 2015.

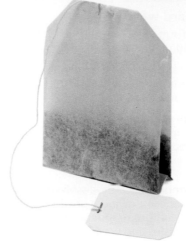

FIGURE 1: Scientists have developed a nanotech teabag. Its nanofibres remove bacteria and contaminants out of water, making the water safe to drink.

What is in tap water?

Salts and microbes

Safe drinking water must not have high concentrations of dissolved salts. It must be free from harmful chemicals and disease-causing microbes.

Only a small amount of Earth's water can be used for drinking.

> Some of the water is locked up in glaciers and icecaps.

> Most of Earth's water is found in the oceans. Unfortunately, the amounts of dissolved salts are so high that drinking it would make you seriously ill.

Adding fluorine

Since the 1960s, fluoride has been added to some toothpastes and water supplies in the UK.

Fluoride salts occur naturally in groundwater, but more than 0.3 parts per million are needed to benefit the health of your teeth. One part per million is ideal.

About 10% of the UK has fluoride concentration at one part per million. In some of these areas, fluoride occurs naturally in the water. In other areas, fluoride is added at the treatment works.

Opinions are divided as to whether all water supplies should have fluoride added.

Arguments in favour of adding fluoride

> It is copying a natural process.

> There have been no harmful outcomes in 60 years.

> Fluoride gives a high degree of protection against tooth decay, especially in children.

Adding chlorine

Domestic water from a mains supply passes through a treatment works.

1. Water is filtered by passing it through filter beds of sand and activated charcoal. This removes any solids.

2. Chlorine gas is added to the water at, typically, one part chlorine per million parts water. This is to sterilise the water.

3. The chlorinated water is stored for up to two hours. It must be long enough to kill most microbes.

4. The water then passes into the distribution system.

FIGURE 2: Why is chlorine added to swimming pools?

Arguments against adding fluoride

> People should be free to drink water without additives.

> The long-term health risks are unknown.

> Too much fluoride can make teeth mottled.

> There is some evidence that fluoride can affect bones (this is not proven).

QUESTIONS

1 70% of Earth's surface is covered by water, yet providing sufficient drinkable water for the world's population is a major challenge. Explain why.

2 Describe how fluoride salts occur naturally in water.

3 Explain why dentists want fluoride added to water supplies.

> **Remember**
> Chlorine is added to public water supplies to disinfect the water. Fluoride salts may be added to improve dental health.

Water filters

Some people choose to purify water by using a water filter. The different types may contain:

> **carbon** (in the form of activated charcoal), which has large pores to absorb impurities as the water passes through

> **ion exchange resins**, which have hydrogen ions, H⁺, bonded to their surfaces – these swap with positively charged ions such as Ca^{2+}, Mg^{2+} and Cu^{2+}, removing the metal ions from in the water

> **silver**, in the form of nano particles, which destroys harmful microbes.

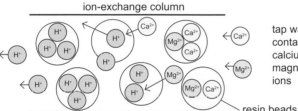

calcium and magnesium ions replaced by hydrogen ions

ion-exchange column

tap water containing calcium and magnesium ions

resin beads

FIGURE 4: Ion exchange happens on the surface of the resin.

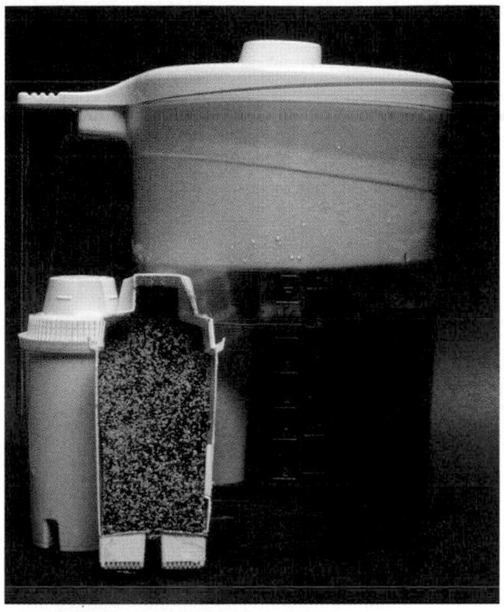

FIGURE 3: This water filter cartridge contains activated charcoal and an ion exchange resin.

QUESTIONS

4 Suggest some advantages and disadvantages of using a home water filter.

5 Explain why exchanged H⁺ ions in water are not an impurity.

Desalination

Desalination is a general term for technologies used to remove dissolved salts from seawater.

One technique is **distillation**, but this requires a lot of energy. Distillation is, therefore, very expensive.

Reverse osmosis is another technique and requires less energy.

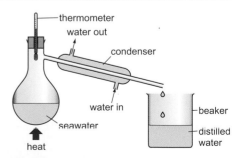

thermometer
water out
condenser
water in
beaker
seawater
distilled water
heat

FIGURE 5: Distillation can be carried out in the lab. Describe the changes of state that happen during distillation.

QUESTIONS

6 Find out about osmosis and how reverse osmosis works.

Preparing for assessment: Applying your knowledge

To achieve a good grade in science, you not only have to know and understand scientific ideas, but you need to be able to apply them to other situations and investigations. These tasks will support you in developing these skills.

✹ Thirsty work

Sam, Sophie and Emma are doing their Duke of Edinburgh's silver award expedition. It is day one of three and the weather forecast is hot and sunny. They have been reminded to take plenty of drink with them, enough to last until they set up camp in the evening. Following the healthy option, they are all packing bottles of water.

Sam is filling her trendy silver coloured drinking bottle with tap water. "I'm not wasting money on bottled water. My tap water has been disinfected at the treatment plant. Some of the chemicals that they added stay in the water, so my water will be safe to drink all day."

Sophie does not like the taste of her tap water and has packed a bottle of natural mineral water. The label says that it is "bottled at source from water flowing deep underground through rocks dating back millions of years". It is checked for harmful microbes and accidental pollutants where it is extracted and bottled there. The label also gives its mineral composition.

Typical mineral analysis	
Mineral	Content (mg/dm³)
calcium	10.0
magnesium	2.5
sodium	9.0
potassium	2.0
hydrogen carbonate	25.0
chloride	12.0
sulfate	10.0
nitrate	11.0

Sophie knows that her natural mineral water must always have the same amounts of minerals in it to be called 'natural mineral water'. It must have nothing else added. That is the law.

Emma is saving for her holiday and has bought a cheaper bottle of plain bottled water. The label just says "Bottled water". Emma knows that this could be from a variety of sources, but the water is tested and treated for harmful microbes. Some brands even have the mineral content adjusted.

✳ Task 1

Sam's tap water has been treated before it reaches her home.

(a) What chemical is added to disinfect tap water?

(b) Sam knows that another chemical has been added to her tap water for health reasons.

(i) Which chemical is this?

(ii) How might it improve her health?

(iii) Some of Sam's neighbours object to having this additive in their tap water.

Give two reasons why they might object.

✳ Task 2

Sophie is happy with her natural mineral water because she can rely on the mineral content.

(a) Which minerals in Sophie's water have health benefits and why?

(b) Explain where the minerals in the water came from.

(c) Sophie thinks that her natural mineral water is temporary hard water.

Why might she think this?

(d) Sam is concerned that Sophie's natural mineral water may not be safe to drink by the end of the afternoon.

What reason does Sam have for her concern?

✳ Task 3

Emma's bottled water may have been treated with UV light to kill harmful microbes. The water could have been filtered through an ion exchange resin to adjust the dissolved mineral content, especially if the water is very hard.

(a) Which metal ions cause hard water?

(b) Explain how ion exchange resins remove these metal ions and make the water soft.

✳ Maximise your grade

Answer includes showing that you…	
E	know that the compounds dissolved in water determine whether it is hard or soft.
	know that some compounds in rocks can dissolve in ground water, causing hard water.
	can name the metal ions responsible for hard water.
C	can distinguish between permanent and temporary hard water.
	can explain how ion exchange resins work to soften water.
A	are aware of the arguments for and against additions to our water supplies.

Energy from reactions

You will find out:

> about exothermic and endothermic reactions

> how calorimetry can be used to measure the energy changes when foods and fuels burn and of reactions in solution

Petrol or pasta?

Cars and people need fuel. Many cars use petrol, many people eat pasta. All fuels react with oxygen. This releases energy that turns the wheels on the car. It releases energy that enables us to move and keep warm. The difference is in how the reactions are started and how they are controlled.

Did you know?

Food scientists use calorimetry to measure the energy available from different foods.

FIGURE 1: Sports drinks contain glucose which is easily absorbed by the body and provides energy.

Energy transfers

Exothermic and endothermic

An **exothermic** reaction releases energy that is transferred to the surroundings, heating them up. The temperature of the surroundings increases.

An **endothermic** reaction absorbs energy that is transferred from the surroundings, cooling them. The temperature of the surroundings decreases.

The amount of energy transferred can be worked out either by experiment or by calculation.

QUESTIONS

1 Describe the difference between an exothermic reaction and an endothermic reaction.

2 Calculate the energy needed to raise the temperature of 50 g of water by 20 °C.

Simple calorimetry

Energy is measured in **joules** (J). It takes 4.2 J to raise the temperature of one gram of water by one degree Celsius. This is the **specific heat capacity** of water and is measured in J/g °C.

When a substance burns, it releases energy. This is an exothermic reaction. The energy released can be transferred to water in a container. By measuring the increase in the temperature of the water, the amount of energy released may be calculated.

This experimental technique is **calorimetry**. It can be used to work out the relative amounts of energy released by different foods and fuels.

You can calculate the energy transferred to water with the equation:

$$Q = m \times c \times \Delta T$$

where

Q = energy transferred to the water (J)
m = mass of water (g)
c = specific heat capacity of water (4.2 J/g °C)
ΔT = change in temperature of the water

Measuring energy transfers

Energy in foods and fuels

The apparatus shown in Figure 2, on the next page, is used to measure the energy given out by candle wax. It can be modified to measure the energy given out by burning nuts and snacks such as corn puffs, and by alcohol fuels.

🔍 ... exothermic AND endothermic reactions

Example

0.20 g of candle wax was burned in the apparatus in Figure 2.
The temperature rise was 35 °C with 25 cm³ water.

The density of water is 1 g/cm³, therefore the mass of water (m) is 25 g.

Using $Q = m \times c \times \Delta T$ for 0.20 g candle wax:

energy transferred $= 25 \times 4.2 \times 35$

$\qquad\qquad\qquad = 3675 \text{ J} = 3.675 \text{ kJ}$

The energy transferred per gram of candle wax $= 3.675 \div 0.20 = 18.38$ kJ/g.

Energy transfers of reactions in solution

Many chemical reactions take place when one or both of the reactants are in solution. Energy changes of these reactions can be measured by mixing the reactants in an insulated container and measuring the temperature change.

Example

100 cm³ of sodium hydroxide solution (1 mol/dm³) was mixed with 100 cm³ of hydrochloric acid (1 mol/dm³). The temperature rise was 6.7 °C.

The mass of the solution was 200 g (assuming its density is 1 g/cm³).

Using $Q = m \times c \times \Delta T$

energy transferred $= 200 \times 4.2 \times 6.7 = 5628$ J or 5.628 kJ

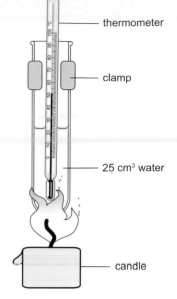

FIGURE 2: Measuring the energy given out by burning candle wax.

⦿ QUESTIONS

3 Burning 0.10 g of corn puffs raises the temperature of 25 cm³ water by 20 °C. Calculate how much energy can be obtained from 1 g of corn puffs.

4 Burning 0.5 g of ethanol raises the temperature of 100 cm³ water by 27 °C. Calculate the energy released when 1 g of ethanol burns.

5 Adding 2.5 g of solid sodium hydroxide to 100 g of water raised the temperature from 22 °C to 28 °C. Calculate the energy transferred.

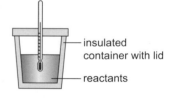

FIGURE 3: Measuring energy transfers of reactions in solution.

Energy values

Although the joule is the standard unit of measurement for energy, **calories** are commonly used, especially by the food industry. A calorie (**cal**) is equal to 4.2 joules, so to convert calories to joules or kcal to kJ, multiply by 4.2.

TABLE 1: Energy content of some breakfast cereals.

Cereal	Energy value (kcal/100g)	Energy value (kJ/100g)
cornflakes	361	1516.2
coco pops	375	1589.0
porridge	150	633.0

TABLE 2: Energy given out by some fuels when 1 mol is burned in oxygen.

Fuel	Energy given out (kJ/mol)
hydrogen	286
natural gas	890
liquid petroleum gas (LPG)	2219
petrol	5470

⦿ QUESTIONS

6 Suggest why energy values for breakfast cereals are measured per 100 g and energy values for fuels are measured per mole.

Energy and bonds

You will find out:

> how energy level diagrams track the energy changes in a reaction

> how bond energies can be used to calculate energy changes

> about activation energy

Forest fire

Whole forests can be destroyed by forest fires. The fires run out control because combustion reactions are highly exothermic. The energy released dries neighbouring trees, making them combust more quickly. As well as a fuel (the trees), oxygen is needed for combustion. Winds provide a plentiful supply. They also spread the flames.

FIGURE 1: Forest fires produce carbon dioxide and water vapour.

Energy level diagrams

In a chemical reaction, old bonds are broken and new bonds are made.

> Energy is needed to break bonds.

> Energy is released when bonds form.

It is not possible to measure the energy stored in a substance. However, the energy transferred in a chemical reaction can be measured. This can be shown on an **energy level diagram**.

Substances react when their particles collide. However, the particles must have sufficient energy to lead to a reaction. This minimum amount of energy is called the **activation energy**. It is why a spark, or other heat source, is needed to start a forest fire.

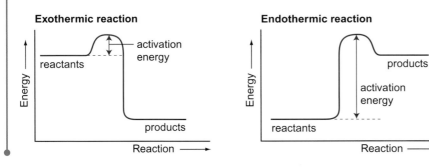

FIGURE 2: Energy level diagrams for an exothermic and an endothermic reaction. The bump on each diagram shows the activation energy.

QUESTIONS

1 Describe the relationship between a successful collision and the activation energy.

2 Explain why fireworks do not react until ignited.

Bond breaking and forming (Higher tier)

Exothermic and endothermic reactions

> In an exothermic reaction, less energy is needed to break bonds in reactants than is released when bonds form to make products.

> In an endothermic reaction, more energy is needed to break bonds in reactants than is released when bonds form to make products.

For example, when methane reacts with oxygen: $CH_4(g) + 2O_2(g) \rightarrow CO_2(g) + 2H_2O(g)$

The reaction is exothermic because the total energy to break four C–H bonds and two O=O bonds is less than the energy released when four O–H bonds and two C=O bonds form.

Bond energy

Bond energy is the amount of energy needed to break one mole of bonds. For example, when one mole of C–H bonds are broken, 413 kJ of energy must be supplied. When one mole of C–H bonds is formed, 413 kJ of energy will be released.

TABLE 1: Some bond energy values.

Bond	H–H	C–H	O=O	O–H	C=O
Bond energy (kJ/mol)	436	413	496	463	743

Calculating energy transfer to surroundings

Example

Calculate the energy transferred to the surroundings when one mole of methane is burned.

1. Energy (kJ) needed to break the bonds in the reactants:

    4 moles of C–H bonds = 4 × 413 = 1652 kJ

    2 moles of O=O bonds = 2 × 496 = 992 kJ

Total amount of energy to break bonds = 2644 kJ

2. Energy released when bonds are made in the products:

    4 moles of O–H bonds = 4 × 463 = 1852 kJ

    2 moles of C=O bonds = 2 × 743 = 1486 kJ

Total amount of energy released = 1852 + 1486 = 3338 kJ

3. More energy is released by new bonds forming than is supplied by the broken bonds. The reaction is exothermic.

The energy released = 3338 – 2644 = 694 kJ

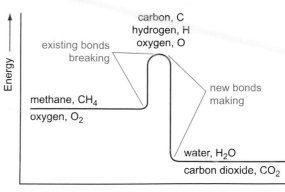

FIGURE 3: Bond breaking.

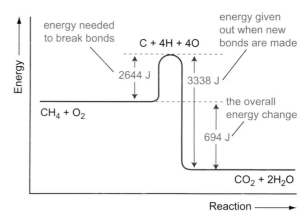

FIGURE 4: Bonds broken and made.

⊙ QUESTIONS

3 Describe the bonds broken and formed when oxygen and hydrogen react to make water.

4 Draw an energy level diagram, similar to Figure 3, for the reaction of oxygen and hydrogen.

5 Using the values in Table 1, calculate the energy change when 1 mole of oxygen reacts with 2 moles of hydrogen to make water.

Average bond energies

Bond energies measure the strength of the bond. Stronger bonds have higher bond energies. The strength of all bonds is affected slightly by other atoms in the molecule.

For example, look at the energies to break the bonds in one mole of water molecules.

Breaking the first O–H bond: H–O–H → H–O + H

needs 502 kJ of energy.

Breaking the second O–H bond: O–H → O + H

needs only 427 kJ.

Average bond energies are calculated using the energy needed to break the bond in a range of different molecules.

⊙ QUESTIONS

6 (i) Calculate the average bond energy for 1 mol O–H bonds in a water molecule.

(ii) Suggest why your answer is different from the value in Table 1.

Energy-saving chemistry

Green chemistry

Green chemistry aims to make the chemical industry more sustainable and environment friendly. New processes and chemical products are needed that reduce or eliminate the use of hazardous substances, produce less waste and use less energy.

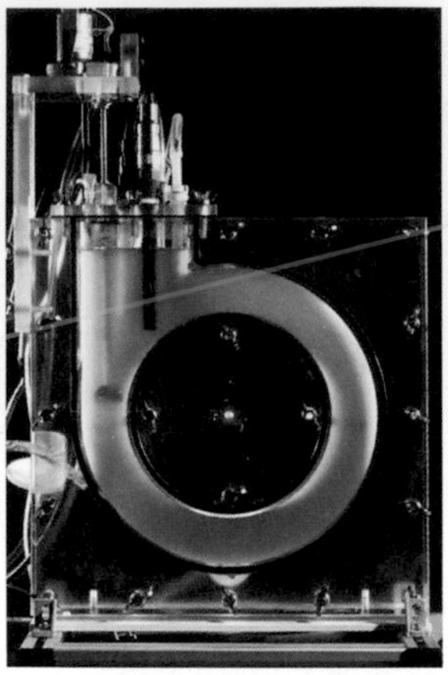

FIGURE 1: Algae in this bioreactor produce hydrogen gas.

You will find out:

> how catalysts speed up chemical reactions

> about the problems of burning hydrogen fuel in a combustion engine

> how hydrogen is used as a fuel in fuel cells

Catalysts reduce energy demands

Catalysts are used in industry to increase the rate of a chemical reaction. Catalysts may take part in the reaction, but can be recovered at the end, chemically unchanged. They may be used many times.

A catalyst works by providing an alternative route for the reaction to happen. This route has a lower activation energy: colliding particles need less energy to have collisions that lead to reaction. The product can be obtained quicker, often using less energy.

Poly(ethene) was discovered by accident in 1933 when ethene was heated to 170 °C at 2000 times atmospheric pressure. Nowadays some poly(ethene) manufacturing processes use titanium and aluminium compounds as catalysts. Polymerisation happens at close to atmospheric pressure and at 60 °C. Less energy is used and the catalyst can be reused.

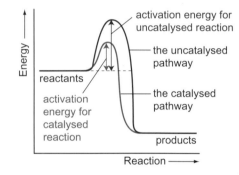

FIGURE 2: Changing the route of a reaction by using a catalyst.

QUESTIONS

1 Explain why catalysts are a green option.

2 Draw an energy level diagram to show a catalysed and uncatalysed endothermic reaction.

Hydrogen – a fuel for the future?

Burning fossil fuels to run cars and heat homes is not sustainable and many scientists are researching alternatives.

Hydrogen as a fuel

Hydrogen reacts exothermically with oxygen to produce water.

$H_2(g) + O_2(g) \rightarrow H_2O(g)$

Hydrogen can be used two ways in a vehicle:

> as a fuel in a combustion engine, replacing petrol

> as a fuel in a fuel cell.

Q ... catalysts AND energy activation

Benefits of hydrogen as a fuel

> No carbon dioxide or sulfur dioxide emissions.

> Burning 1 g of hydrogen releases 143 kJ of energy.
Burning 1 g of petrol releases 48 kJ.

Disadvantages

> Liquid hydrogen has one-tenth the density of petrol. Therefore, vehicles need large fuel tanks to carry enough fuel for a car to travel as far on hydrogen as it can on petrol.

> Fuel tanks must carry compressed hydrogen safely.

> Few people are likely to buy hydrogen-powered cars until most filling stations have hydrogen pumps.

Combustion engines

It is relatively simple to convert existing cars to run on hydrogen. However, only 20% of the energy from combustion is converted to mechanical energy to turn the wheels. The rest is 'wasted', mostly by heating up the surroundings.

If 286 kJ are released when one mole of hydrogen combusts in a combustion engine, then only 20% of 286 kJ – or 57.2 kJ – are available to move the car. The rest makes the car engine extremely hot and is transferred eventually to the air around the car.

Fuel cells

Two reactants are fed into a fuel cell. The cell does not run out or need recharging as long as the reactants are supplied. The chemistry inside the fuel cell produces an electric current that can be used to power vehicles, heat buildings and provide electricity.

Each fuel cell produces a potential difference of between 0.4 and 1.0 volts. Several fuel cells can be wired in series to give higher potential differences.

Fuel cells do not waste as much energy as the combustion engine. They are up to 60% efficient.

FIGURE 3: This car is fuelled by hydrogen gas. Why is hydrogen called 'clean energy'?

Did you know?

Some scientists believe hydrogen, the most abundant element in the Universe, will be the green fuel of the future.

QUESTIONS

3 Describe the problems that scientists need to solve, before hydrogen fuel can replace petrol in cars.

4 It is 'greener' to use hydrogen in a fuel cell, rather than burn it in a combustion engine, to power a vehicle. Explain why.

How does a fuel cell work?

Hydrogen gas is fed to the anode and oxygen gas to the cathode. The platinum catalyst separates the electrons from the protons of the hydrogen atoms. The separated electrons are fed round a circuit from the anode to the cathode, producing an electric current.

QUESTIONS

5 Explain why scientists are trying to find alternative, cheaper, materials that will act as catalysts in fuel cells.

6 Find out about the latest developments in fuel cell technology.

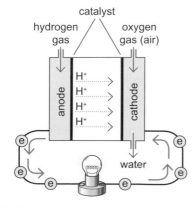

FIGURE 4: Hydrogen fuel cell. The platinum, used as the catalyst, is expensive and rare.

Preparing for assessment: Planning an investigation

To achieve a good grade in science, you not only have to know and understand scientific ideas, but you need to be able to apply them to other situations and investigations. These tasks will support you in developing these skills.

✳ The energy value of food

School dinners are changing for the better. It is out with the junk food and in with more nutritious food

A school canteen manager wants to persuade more students to eat the healthy meals on offer. The manager has suggested that students use their science lessons to investigate the nutrition and energy available from different foods and how this compares with our daily requirements.

In chemistry lessons, students investigate how much energy is released when different foods are burned. This is the energy available when you eat them. The students compare crisps with brown bread, pasta and cheese.

How would you carry out this investigation?

✳ Researching

1. Carry out research to find a method for an experiment. Check that it is suitable to use in class. What apparatus is available?

2. The energy contents of foods are commonly measured in laboratories and often printed on the sides of packaging.

How much information about energy values is given on the sides of relevant food packets? How can you use these data to compare with your own results?

Data gathered from existing sources, such as the internet and textbooks, are secondary data. The internet will give you a safe experiment, if you select good websites.

The units used may confuse. You can expect to come across joules, calories and even BTUs. Keep to the units used in science lessons.

Beware of information on students' own websites and the ask-a-question type website.

Values are often labelled *Nutritional facts*. Check the units used. Energy content may be quoted for a typical serving or for 100 g. You will have to calculate the energy content for one gram.

Manufacturer's values are likely to be more accurate than school experiment results.

✳ Connections

How Science Works
- Plan practical ways to develop and test scientific ideas
- Assess and manage risks when carrying out practical work
- Collect primary and secondary data
- Select and process primary and secondary data

Science ideas
C3.3.1 Energy from reactions

✳ Planning

3. Having chosen a method, write up how you are going to do the experiment, step by step, to obtain your primary data.

(a) Identify the independent and dependent variables.

(b) Your test must be fair. List all the variables that you need to keep the same for each test.

(c) List all the measurements that you need to make and the units that you will use.

(d) Do you need to repeat each test and if so, how many times?

(e) How will you use your results? How will you compare the energy from each food sample?

Primary data is information that you collect from your investigation.

The independent variable is the factor that you change in your experiment. The dependent variable is what changes as a result.

Controlling variables makes your investigation valid.

Think about the calculations that you will need to do.

✳ Assessing and managing risks

4. How will you keep your workplace and procedure safe for yourself and others? Itemise all the potential hazards in a risk assessment.

Repeats improve the reproducibility of an experiment.

This should identify each risk and the steps needed to minimise it.

✳ Collecting primary data

5. You will need a table to record your results. The table can include measurements you take only, or calculations as well. Each column must have a correct heading with units.

6. Two students used the apparatus, shown right, to carry out their investigation.

These are their results for a sample of potato crisps. They repeated the test on the crisps three times.

Test 1: 0.50 g crisps raised the temperature of 25 cm³ water from 18 °C to 48 °C

Test 2: 0.45 g crisps raised the temperature of 25 cm³ water from 15 °C to 50 °C

Test 3: 0.50 g crisps raised the temperature of 25 cm³ water from 16 °C to 35 °C

(a) Make a results table to show these results. Include temperature change (°C), energy given out by each sample (J) and energy given out by 1 g of each sample (J).

(b) Calculate a mean value for the energy given out when 1 g of crisps is burned.

(c) Indicate how you would extend your table to include results for bread, pasta and cheese.

7. Do you consider that the results are valid, accurate and reproducible?

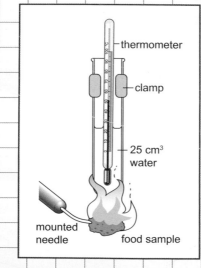

The method used in this example allows a lot of heat to be lost from the burning food. The accuracy of the results will be low.

Checklist C3.1–3.3

To achieve your forecast grade in the exam you will need to revise

Use this checklist to see what you can do now. Refer back to the relevant topics in this book if you are not sure. Look across the three columns to see how you can progress. Bold text means Higher tier only.

Remember that you will need to be able to use these ideas in various ways, such as:

> interpreting pictures, diagrams and graphs

> applying ideas to new situations

> explaining ethical implications

> suggesting some benefits and risks to society

> drawing conclusions from evidence you are given.

Look at pages 188–209 for more information about exams and how you will be assessed.

To aim for a grade E	To aim for a grade C	To aim for a grade A
Know that Newlands and Mendeleev devised early periodic tables.	Know that early periodic tables listed the elements in order of atomic weights. Know why Mendeleev left gaps in his periodic table and reversed the order of some elements.	
Know that Mendeleev's periodic table developed into the modern periodic table used today.	Know that the modern periodic table lists elements in order of increasing atomic number and places elements in appropriate groups. Describe the electron configuration of an atom from its position in the periodic table.	
Name some elements in Group 1. Describe some of the properties of Group 1 elements, using word equations.	Interpret data on Group 1 elements, describe trends within the group. Use word equations to describe their chemical reactions.	Use balanced symbol equations to describe reactions of Group 1 elements with non-metals and with water.
Recall examples of transition metals and some of their properties.	Interpret transition metal data. Describe the properties of transition metals.	
Recall examples of Group 7 elements and describe some of their properties.	Know the trends in melting point, boiling point and reactivity in Group 7 elements.	**Explain reactivity trends in terms of electrons and energy levels.**
Use word equations to describe displacement reactions.		Use symbol equations to describe displacement reactions.

To aim for a grade E   To aim for a grade C   To aim for a grade A

To aim for a grade E	To aim for a grade C	To aim for a grade A
Know that the compounds dissolved in water determine whether water is hard or soft.	Know that the calcium or magnesium compounds in hard water dissolved from rocks.	Write balanced symbol equations for the formation of hard water (temporary).
Know how soap reacts with hard and soft water.	Measure the hardness of water by titration with soap solution Appreciate the economic aspects of water hardness.	
Know the effects of limescale in hard water areas.	Know the effect of boiling on temporary and permanent hard water.	**Write ionic equations for the decomposition of hydrogencarbonate ions in temporary hard water when it is heated.**
Know that washing soda and water filters can soften hard water.	Describe how ion exchange resins replace calcium and magnesium ions in hard water.	Write balanced symbol equations to show how washing soda softens water.
Understand why tap water may have chlorine or fluoride salts added.	Be aware of the arguments for and against the addition of chlorine and fluoride salts to drinking water.	
Know that distillation can be used to produce pure water.	Understand the economic implications of producing pure water by distillation.	
Know that combustion of food and fuels is exothermic. Know that the energy change can be measured.	Use simple calorimetry to measure and calculate the energy change in combustion reactions, reactions with water and neutralisation reactions. Interpret energy change data and convert units where required.	
Recognise energy level diagrams as exothermic or endothermic reactions.	Know how activation energy and the effect of a catalyst is shown on an energy level diagram.	
Know that bonds are broken and made in a chemical reaction.	Know that energy must be supplied to break bonds and is given out when new bonds are made.	**Compare the energies absorbed and released in bond breaking and making for exothermic and endothermic reactions.** **Use bond energies to calculate the energy transferred in reactions.**

In the examination, data and the periodic table will be given on a separate sheet. You will be expected to select appropriate data from the sheet.

1. In the 1800s, many new elements were discovered. Mendeleev tried to make sense of the new data by sorting elements into a periodic table.

AO1 **(a)** Mendeleev listed the known elements in order. What property did he use to make this list? [1]

AO1 **(b)** Mendeleev lined up the elements with similar properties in columns. What are these columns called? [1]

AO1 **(c)** Mendeleev left gaps in his periodic table. Explain why. [1]

AO1 **(d)** What property is used to put the elements in order in the modern periodic table? [1]

2. A group of students analysed the tap water in their area. They found that:

• soap forms a scum with the water before it lathers

• kettles and water pipes are coated inside with limescale.

They decided that the water is hard.

The students tested different water samples. They measured the volume of soap solution needed to produce a permanent lather. They used 10 cm³ of each sample. They also tested pure deionised water for comparison. The table shows their results.

Sample number	Water sample	Volume of soap solution (cm³)
1	pure deionised water	2
2	untreated tap water	15
3	boiled tap water	5
4	tap water passed through water filter	8

AO3 **(a)** Which of the water samples is the softest? [1]

AO3 **(b)** What type of hard water is present? Explain your answer and give the evidence. [3]

In part (c) of this question you will be assessed on using good English, organising information clearly and using specialist terms where appropriate.

AO1 **(c)** Passing hard water through a water filter had a similar result to boiling.

Name two substances commonly found in water filters and explain what they do. [5]

3. Calorimetry experiments measure the energy released when ethanol burns.

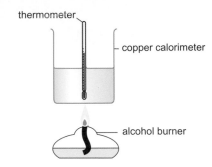

Students measured the mass of the ethanol burner before and after the experiment.

They used 100 cm³ (100 g) water in the copper can and measured its initial and final temperature.

Their results are shown in the table.

Mass of burner (g)		Temperature of water (°C)	
initial	final	initial	final
55. 5	54.8	15.0	30.5

AO2 **(a)** Calculate the mass of ethanol burned. [1]

AO2 **(b)** What was the temperature rise? [1]

AO2 **(c)** How much energy was released when the ethanol burned? [2]

Use the equation $Q = m \times c \times \Delta T$ where

Q is energy released in joules

m is mass of water in grams

C is specific heat capacity of water (4.2 J/g °C)

ΔT is the change in temperature.

AO2 **(d)** Calculate the energy given out when one gram of ethanol burns. [2]

AO3 **(e)** An online database gives the accurate value for burning 1 g of ethanol as 20.15 kJ. Give **two** reasons why the students' result is lower. [2]

AO1 recall the science   AO2 apply your knowledge   AO3 evaluate and analyse the evidence

✴ WORKED EXAMPLE – Foundation tier

The motor industry now makes several vehicles that use fuel cells. The fuel cell generates electricity that powers the car. Electricity is produced when hydrogen gas reacts with oxygen in the fuel cells.

(a) Write a word equation for the reaction in the fuel cell. [1]

hydrogen + oxygen = water

(b) The energy diagram below follows the reaction in the hydrogen fuel cell.

Label the diagram to show:

(i) the overall energy change

(ii) the activation energy

(iii) the energy given out when new bonds form in water. [3]

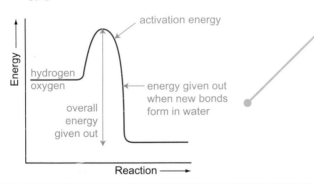

In part (c) of this question you will be assessed on using good English, organising information clearly and using specialist terms where appropriate.

(c) Evaluate the advantages and disadvantages of using hydrogen as a fuel for cars in the future. [4]

Petrol is running out but we have lots of hydrogen and it's free. Fuel cells make water which doesn't pollute like petrol. So it is better for the environment and doesn't harm animals and plants or us. The air is better to breathe for people with bad health.

How to raise your grade!
Take note of these comments – they will help you to raise your grade.

Never use an equals sign in a chemical equation. You must only use an arrow or the equilibrium sign. This should be an arrow. The reactants and products are correct.

This answer correctly shows an amount of energy for all three labels, with a double-headed arrow.

Note that this question is about an exothermic reaction. Make sure that you can label similar diagrams for an endothermic reaction.

This answer shows some knowledge and understanding, such as fuel cells producing water and being non-polluting. However, hydrogen is not free. There is plenty of water that can be used to produce hydrogen, but the process uses electricity which is not free. The point about air quality is valid.

The answer has some structure and organisation but uses only a few specialist terms. 'Fossil fuel' and 'finite resource' could be included.

No problems, such as the availability of hydrogen at fuel stations, were included.

There are no spelling errors, but there are a few grammatical errors. This answer would gain two marks.

In the examination, data and the periodic table will be given on a separate sheet. You will be expected to select appropriate data from the sheet.

1. Halogen bulbs have a tungsten filament and are filled with a mixture of an inert gas and small amounts of bromine (or iodine). When the bulb is on, the tungsten (symbol W) filament reacts with bromine to make tungsten bromide.

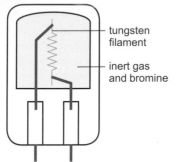

tungsten filament

inert gas and bromine

AO1 **(a) (i)** What is the charge on the bromide ion? [1]

AO2 **(ii)** Write a balanced symbol equation for the reaction between tungsten and bromine. The formula of tungsten bromide is WBr_6. [1]

(b) Manufacturers do not use fluorine in halogen bulbs because it is too reactive.

AO1 **(i)** Use ideas about the energy levels of electrons in an atom, to explain why fluorine is more reactive than bromine. [3]

AO1 **(ii)** The reactivity of the halogens decreases down the group. What is the trend in reactivity for Group 1 metals? [1]

AO2 **(c)** Tungsten is a transition metal. Its melting point is 3422 °C. It does not melt in halogen bulbs.

Give two reasons why the Group 1 metal sodium is not suitable for making halogen bulb filaments. [2]

AO2 **(d)** Tungsten can form ions with a 2+ charge and with a 3+ charge. Suggest the correct formulae for two bromides of tungsten. [2]

AO3 **2.** *In this question you will be assessed on using good English, organising information clearly and using specialist terms where appropriate.*

Some water companies add fluorine to their water supplies and others do not. Explain why fluorine is added to water supplies and describe two advantages and two disadvantages of adding fluorine to the water supply. [6]

3. Hydrogen gas and chlorine gas do not react in the dark. In strong sunlight, they react explosively to produce hydrogen chloride gas. This strongly exothermic reaction has caused many accidents.

AO1 **(a)** What does 'exothermic' mean? [1]

AO1 **(b)** Energy from sunlight is needed to start the reaction. What is this energy called? [1]

AO2 **(c)** Show this reaction on an energy level diagram.

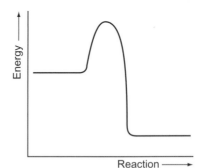

Reaction ⟶

Label the following:
- hydrogen and chlorine reactants
- hydrogen chloride product
- energy needed to break existing bonds
- energy given out when new bonds form
- the overall energy change. [5]

AO2 **(d)** Use the following information in the questions.

The equation for the reaction between hydrogen and chlorine is

$H_2(g) + Cl_2(g) \rightarrow 2HCl(g)$

Bond energies	
Bond	**Bond energy (kJ/mol)**
H–H	436
Cl–Cl	242
H–Cl	431

(i) Calculate the total energy needed to break one mole of H–H bonds and one mole of Cl–Cl bonds. [1]

(ii) Calculate the energy given out when two moles of hydrogen chloride bonds form. [1]

(iii) Use your answers to (i) and (ii) to calculate the overall energy change for the reaction. [2]

✳ WORKED EXAMPLE – Higher tier

The diagram shows a small section of today's periodic table.

Group 6	Group 7	Group 0
128	127	130
Te	I	Xe
Tellurium	Iodine	Xenon
52	53	54

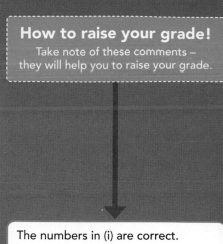

How to raise your grade!
Take note of these comments –
they will help you to raise your grade.

(a) (i) Which of the numbers in the boxes were not known to Mendeleev? [1]

52, 53 and 54

(ii) Explain why they were not known to Mendeleev. [1]

They hadn't been discovered yet.

The numbers in (i) are correct. However, to obtain the second mark, for part (ii), the candidate needed to say that these are atomic numbers and were not used when Mendeleev composed his periodic table.

(b) Iodine and tellurium appeared in the same order in Mendeleev's periodic table. Why was this order unusual in Mendeleev's periodic table? [1]

Mendeleev normally put the elements in order of atomic weight in his periodic table. The atomic weight of tellurium is higher than the atomic weight of iodine.

This is clearly explained and is a good answer.

(c) Explain why Mendeleev chose to put tellurium and iodine in this order. [1]

Because the properties matched the groups.

This answer is minimal and would not receive the mark.

(d) Which of the elements in the table above, did not appear in Mendeleev's early periodic table and why? [2]

Xenon, because the noble gases had not been discovered at this time.

A good answer is given here.

The noble gases had not been discovered because they are very unreactive.

(e) Group 7 elements react with Group 1 elements to form halides.

(i) Write a balanced symbol equation for the reaction between sodium and iodine. [1]

$2Na + I_2 \rightarrow 2NaI$

(ii) Explain why the reaction between potassium and iodine is more violent than the reaction between lithium and iodine. [3]

These are both reactions of Group 1 metals with iodine, but potassium is more reactive than lithium. Group 1 metals are more reactive as you go down the group, so potassium reacts more violently with iodine than lithium.

There are three marks for this question. This answer makes only one point – that reactivity increases down Group 1. The candidate gains only one mark.

The candidate needed to explain the reactivity in terms of atomic structure: potassium atoms are larger and have more filled shells of electrons (second mark). This decreases the attraction of the nucleus to the outer electron and the outer electron is more easily lost (third mark).

Chemistry C3.4–3.6

What you should know

Acids, bases and salts

An acid reacts with a base to form a salt and water in a neutralisation reaction.

When exact reacting amounts of acid and base are added, the pH will be 7.

Salts contain a metal ion and a non-metal ion.

 Name an acid, and an alkali, and the salt that they make in a neutralisation reaction.

Rates of reaction

Temperature, concentration, pressure, particle size and catalysts affect the rate of a reaction.

The collision theory explains how chemical reactions happen and how quickly they happen.

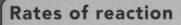 **How do changes in temperature and concentration affect the rate of a reaction?**

Reversible reactions

A reversible reaction can go in either direction and the reaction is never complete.

The yield from a chemical reaction can be calculated from the theoretical yield and the actual yield.

 How do you show that a reaction is reversible in a chemical equation?

Organic chemistry

A family of organic chemicals with the same functional group form a homologous series.

Structural formulae show how the atoms are bonded together.

Name and give the chemical formula for the first three alkanes (a homologous series).

C_2H_5OH

You will find out

Further analysis and quantitative chemistry

> Flame tests and sodium hydroxide solution can be used to detect some metal ions.

> Chemical tests identify non-metal ions.

> One mole of any substance contains the same number of particles, and this is called Avogadro's number.

> Acid–alkali titrations can be used to find unknown concentrations.

The production of ammonia

> A reversible reaction at equilibrium contains constant amounts of reactants and products.

> Changing the temperature changes the yields in a reversible reaction.

> Changing the pressure can change the yield in reversible reactions involving gases.

> The Haber process is used to manufacture ammonia.

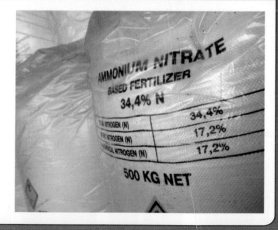

Alcohols, carboxylic acids and esters

> Alcohols and carboxylic acids are both homologous series.

> Compounds in each series have similarities and trends in their structure, properties and uses.

> An alcohol and carboxylic acid react together to form an ester.

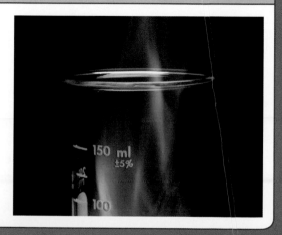

Analysis – metal ions

You will find out:
> how to do a flame test
> how to identify a metal ion using a flame test
> how to identify a metal ion from its hydroxide

Coloured candle flames

Traditional candles burn wax and have mostly yellow flames. Some novelty candles contain small amounts of metal salts. When the candles burn, their flames have different colours depending on which metal salt is present.

FIGURE 1: Why do these candles burn with coloured flames?

Tests to identify metal ions

Flame test

Flame tests are used to identify metal ions in salts. Often, this is the easiest way to identify a Group 1 or Group 2 metal.

To do a flame test you need a clean flame test wire (platinum, preferably, or nichrome).

> Clean the wire: repeatedly dip it in concentrated hydrochloric acid (with care) and then heat the wire in a blue Bunsen flame until no colour is seen.

> Dip the cold wire into concentrated hydrochloric acid and then a sample of the salt.

> Hold the wire in a blue Bunsen flame. Note the colour of the flame.

Some flame test colours are more intense than others. The yellow flame from sodium ions is very intense and persistent – often, it masks colours from other metal ions present.

Remember

Flame tests and addition of sodium hydroxide can be used to identify some metal ions.

nichrome wire

watchglass

sample

FIGURE 2: What is used to clean the nichrome wire, before a flame test?

Lithium ions give a crimson flame.

Sodium ions give a yellow flame.

Potassium ions give a lilac flame.

Calcium ions give a red flame.

Barium ions give a green flame.

Figure 3: Flame test colours.

Sodium hydroxide test

Some metal ions can be identified from their reaction with sodium hydroxide.

The sample must be dissolved in water and then a few drops of dilute sodium hydroxide solution added. The hydroxide formed is often insoluble and appears as a **precipitate** in the solution.

The colour of the precipitate can be used to identify the metal ions present.

QUESTIONS

1 An unknown salt gives a red flame test and makes a white precipitate with sodium hydroxide solution. The precipitate does not dissolve in excess sodium hydroxide solution.
Which metal is present?

2 Samples of iron(II) chloride and iron(III) chloride have been muddled. How can you identify them?

3 Make a risk assessment for carrying out flame tests.

Q ... flame tests

TABLE 1: Observations when sodium hydroxide solution is added to metal ions.

Ion	aluminium, $Al^{3+}(aq)$	calcium, $Ca^{2+}(aq)$	magnesium, $Mg^{2+}(aq)$	copper(II), $Cu^{2+}(aq)$	iron(II), $Fe^{2+}(aq)$	iron(III), $Fe^{3+}(aq)$
Observation	white precipitate of $Al(OH)_3$	white precipitate of $Ca(OH)_2$	white precipitate of $Mg(OH)_2$	blue precipitate of $Cu(OH)_2$	green precipitate of $Fe(OH)_2$	brown precipitate of $Fe(OH)_3$
	precipitate dissolves in excess NaOH(aq)	precipitate does not dissolve in excess NaOH(aq)	precipitate does not dissolve in excess NaOH(aq)			

Forming a precipitate

The solubility of the hydroxides follows a pattern.

> Group 1 hydroxides are soluble.

> Most Group 2 hydroxides are slightly soluble.

> Transition metal hydroxides and aluminium hydroxide are insoluble.

When sodium hydroxide solution is added to a salt solution (the salt has to be soluble), a mixture of all the ions is obtained. If an insoluble hydroxide can form, it will precipitate out. For example:

$$Cu^{2+}(aq) + 2OH^{2-}(aq) \rightarrow Cu(OH)_2(s)$$

Copper(II) hydroxide is insoluble and forms as a blue precipitate.

Since transition metal ions are usually coloured, you can identify the metal present from the colour of the precipitate.

QUESTIONS

4 Why is it not possible to identify potassium ions by adding sodium hydroxide solution?

5 Write a symbol equation for the reaction between
(a) iron(II) sulfate solution and sodium hydroxide solution
(b) iron(III) sulfate solution and sodium hydroxide solution.

Sophisticated flame tests

More often than not, scientists want to work out how much of a substance is present, not simply identify the substance. One instrument used is a flame photometer. It measures the intensity of light given off in a flame test.

More sophisticated techniques use atomic emission spectroscopy or atomic absorption spectroscopy, but both are based on the idea of flame tests.

QUESTIONS

6 Find out how a flame photometer can be used to determine the concentration of a metal ion in a solution.

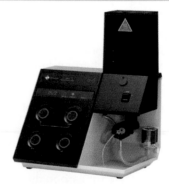

FIGURE 4: A flame photometer is used to measure the concentrations of metal ions in solution.

Q ... solubility hydroxides

Analysis – non-metal ions

You will find out:

> how to identify carbonates using dilute hydrochloric acid
> how to identify halides using silver nitrate solution
> how to identify sulfate ions using barium chloride solution

Monitoring the sea

Marine scientists constantly monitor the composition of the sea, checking for pollution and possible effects of global warming. The most common non-metal ion in seawater is the chloride ion, closely followed by the sulfate ion. Their concentration has stayed constant for millions of years. Concentrations of other non-metal ions, such as the carbonate ion, have varied.

FIGURE 1: Taking samples of seawater for analysis.

Tests to identify non-metal ions

Halide ions

A sample of the unknown compound must be dissolved in water.

Add a few drops of dilute nitric acid, followed by a few drops of silver nitrate solution.

You can detect if a halide ion such as chloride, bromide or iodide is present, if a precipitate forms. Its colour can be used to identify the halide:

> chloride ions – white precipitate of silver chloride

> bromide ions – cream precipitate of silver bromide

> iodide ions – yellow precipitate of silver iodide

Sulfates

The sample must be dissolved in water.

Add a few drops of dilute hydrochloric acid, followed by a few drops of barium chloride solution.

If a sulfate is present, the test will give a white precipitate of barium sulfate.

Carbonates

The solid or, if soluble, a solution can be tested.

A sample containing carbonate fizzes when dilute hydrochloric acid is added slowly. The gas given off is carbon dioxide.

You can test for carbon dioxide using a drop of limewater on the end of a glass rod. The limewater will turn cloudy as a white precipitate of calcium carbonate forms.

FIGURE 2: Silver chloride is white, silver bromide is cream, and silver iodide is yellow.

QUESTIONS

1 Describe how you could show that seashells contain a carbonate.

2 A sample of seawater was divided into two. One was tested with silver nitrate solution and the other with barium chloride solution. Each test produced a white precipitate. What conclusion can you draw about the seawater sample?

3 Describe how you can distinguish between a chloride, a bromide and an iodide.

Reactions in the tests

Halide test

Most chlorides, bromides and iodides are soluble, but the silver halides are insoluble. If a silver halide is made in a reaction, it forms as a precipitate, indicating that a halide is present. The equations for the reactions of solutions of sodium halides with silver nitrate solution are given below.

$$NaCl(aq) + AgNO_3(aq) \rightarrow AgCl(s) + NaNO_3(aq)$$
white precipitate

$$NaBr(aq) + AgNO_3(aq) \rightarrow AgBr(s) + NaNO_3(aq)$$
cream precipitate

$$NaI(aq) + AgNO_3(aq) \rightarrow AgI(s) + NaNO_3(aq)$$
yellow precipitate

Silver nitrate solution contains silver ions, $Ag^+(aq)$, and nitrate ions, $NO_3^-(aq)$. In solution, halide ions, $X^-(aq)$, react with silver ions in silver nitrate solution to form insoluble silver halides, $AgX(s)$. There is a general ionic equation for these reactions.

$$X^-(aq) + Ag^+(aq) \rightarrow AgX(s)$$

Carbonate test

All carbonates give off carbon dioxide with dilute acids. For example:

$$ZnCO_3(aq) + 2HCl(aq) \rightarrow ZnCl_2(aq) + H_2O(\ell) + CO_2(g)$$

When the carbon dioxide is bubbled through limewater (calcium hydroxide solution), calcium carbonate forms as a white precipitate. This confirms the presence of carbonate ions in the original substance.

$$Ca(OH)_2(aq) + CO_2(g) \rightarrow CaCO_3(s) + H_2O(\ell)$$
white precipitate

Sulfate test

Most sulfates are soluble, but barium sulfate is an exception. Here is the equation for the reaction of sodium sulfate solution with barium chloride solution.

$$Na_2SO_4(aq) + BaCl_2(aq) \rightarrow BaSO_4(s) + 2NaCl(aq)$$
white precipitate

Barium chloride solution contains barium ions, $Ba^{2+}(aq)$, and chloride ions, $Cl^-(aq)$. This is the general ionic equation for the reaction of barium ions with sulfate ions in solution.

$$SO_4^{2-}(aq) + Ba^{2+}(aq) \rightarrow BaSO_4(s)$$

FIGURE 3: A white precipitate is a positive result for a sulfate test.

QUESTIONS

4 Write a balanced symbol equation for (a) magnesium chloride solution reacting with silver nitrate solution (b) copper(II) carbonate reacting with dilute sulfuric acid.

Toxic, but useful

Barium salts, including barium sulfate, are toxic. Yet barium sulfate is used in barium meals to diagnose diseases of the digestive system.

X-rays pass through the soft tissues of the digestive tract, but if the tissues are coated with a barium salt, X-rays are stopped and an image can be taken. The patient swallows the barium meal, a mixture containing barium sulfate, and a few minutes later the X-ray is taken.

The insolubility of barium sulfate explains why a toxic salt can be ingested. It is so insoluble that little or none dissolves into the body fluids. The barium sulfate therefore passes through the body.

QUESTIONS

5 Find out what instructions are given to patients before they have a barium meal.

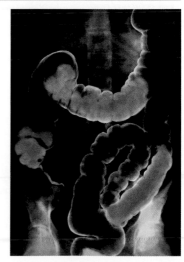

FIGURE 4: An X-ray of the large intestine.

Measuring in moles

You will find out:
> how to calculate relative formula masses
> what a mole is
> how to convert moles to masses and masses to moles
> why moles are useful in concentrations and equations

Bits and bytes

A bit in the hard drive of your computer holds two pieces of information, 0 and 1. Bits are so small, it is not practical to work on all of them at a time, so they are organised into bigger groups called bytes. A byte is a group of eight bits and holds enough information to store a single letter such as 'h'. Scientists working with atoms and molecules face a similar problem, but on a larger scale. The particles involved in chemical reactions are too small to count.

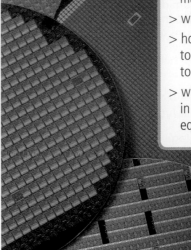

FIGURE 1: Silicon chips on silicon wafers. Each silicon atom has a diameter of 2.71×10^{-10} m.

Masses and moles

Relative formula mass

Chemists often use the term **relative formula mass** to compare chemical substances.

To calculate a substance's relative formula mass, add up the relative atomic masses (A_r) of all the atoms shown in its formula.

A_r values from Table 1 are used in the examples in Table 2.

TABLE 2: Relative formula mass.

Substance	Structure	Formula	Relative formula mass
diamond	giant	C	12
sodium chloride	giant	NaCl	$23 + 35.5 = 58.5$
silicon dioxide	giant	SiO_2	$28 + (2 \times 16) = 60$
oxygen gas	molecular	O_2	$(2 \times 16) = 32$
methane	molecular	CH_4	$12 + (4 \times 1) = 16$

TABLE 1: Some elements and their A_r values.

Element	A_r
H	1
C	12
O	16
Na	23
Mg	24
Si	28
S	32
Cl	35.5
Ca	40
Fe	56

Moles

Scientists count particles in **moles** (abbreviated to **mol**). One mole of any substance is the relative formula mass in grams.

One mole of carbon has a mass of 12 g and one mole of sodium chloride has a mass of 58.8 g.

One mole of any substance contains the same number of particles – that is, 6.02×10^{23}. This is called **Avogadro's number**.

> One mole of carbon atoms contains 6.02×10^{23} atoms and has a mass of 12 g.

> One mole of sodium chloride contains 6.02×10^{23} formula units and has a mass of 58.5 g.

QUESTIONS

1 Calculate the relative formula mass of
(a) iron, Fe
(b) carbon dioxide, CO_2
(c) ethanol, C_2H_5OH
(d) sodium sulfate, Na_2SO_4.

Q ... mass AND moles

Conversions

Converting mass to moles

You can convert a mass of any substance to moles using:

moles = mass (g) ÷ relative formula mass

Example 1

How many moles are there in 60 g of sodium hydroxide, NaOH?

(A_r values: Na = 23, O = 16, H = 1)

1. Calculate the relative formula mass of sodium hydroxide

23 + 16 + 1 = 40

2. Substitute the relative formula mass into the equation

moles = 60 ÷ 40 = 1.5

Converting moles to mass

You can convert moles to mass using:

mass (g) = moles × relative formula mass

Example 2

What is the mass in grams of 0.05 mol calcium hydroxide, $Ca(OH)_2$?

(A_r values: Ca = 40, O = 16, H = 1)

1. Calculate the relative formula mass of calcium hydroxide

40 + (2 × 16) + (2 × 1) = 74

2. Substitute the relative formula mass into the equation

mass = 0.05 × 74 = 3.7 g

Remember

A mole is the relative formula mass in grams.

Did you know?

Scientist count particles in groups of 602 000 000 000 000 000 000 000 (or $6.02 × 10^{23}$). This number of particles is a 'mole of particles'. The particles might be atoms, ions or molecules.

QUESTIONS

2 How many moles are there in
(a) 64 g oxygen
(b) 120 g copper(II) sulfate, $CuSO_4$
(c) 10 g calcium carbonate, $CaCO_3$?

3 Find the mass of
(a) 5 mol sulfuric acid, H_2SO_4
(b) 0.1 mol magnesium sulfate, $MgSO_4$
(c) 0.25 mol methane, CH_4.

Concentration (Higher tier)

The amount of solute dissolved in a solution is its concentration. Concentration can be measured two ways:

> in grams per cubic decimetre of solution (g/dm^3)

> in moles per cubic decimetre of solution (mol/dm^3)

For example, a 1 mol/dm^3 solution of sodium hydroxide contains one mole of sodium hydroxide in one cubic decimetre of solution. Since 1 mol NaOH contains 40 g, there are 40 g of sodium hydroxide in 1 dm^3 solution.

QUESTIONS

4 Calculate the mass of sulfuric acid in 1 dm^3 of a 1 mol/dm^3 solution.

5 Calculate the mass of sodium hydroxide in 100 cm^3 of a 0.1 mol/dm^3 solution.

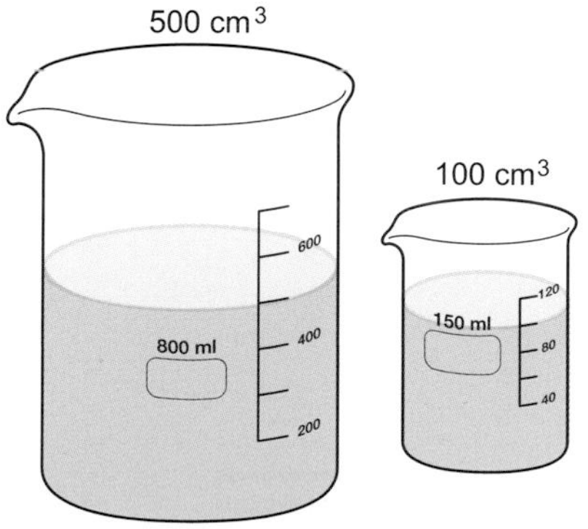

500 cm³

100 cm³

FIGURE 2: Both beakers contain 1 mol/dm^3 sodium hydroxide solution. How many grams of sodium hydroxide are in each volumetric flask?

🔍 ... molarity

Analysis – acids and alkalis

You will find out:

> how to carry out a titration

> how to use titration results to calculate the concentration of one of the reactants

Keeping the same face

Cosmetics must be carefully tested before leaving the factory. Products such as lipstick, moisturiser and body lotion are basically mixtures of water and oil, with added ingredients. Scientists use titrations to check the percentage of water in these products.

FIGURE 1: Too much water would make the lipstick too soft.

Acid–alkali titrations

Scientists use **titrations** to measure the volumes of acidic and alkaline solutions that react with one another. The equipment for a titration includes:

> a suitable **indicator** – to show when the reaction is complete

> **pipette** – to measure out the volume of acid

> **burette** – to measure the volume of alkali with which the acid reacts.

Indicators

TABLE 1: Acid–base indicators and their colours.

Indicator	Colour		
	in acid solution	at end point	in alkaline solution
universal indicator	red	green	violet
phenolphthalein	colourless	pink (persists)	pink

QUESTIONS

1 Why is the 0 cm^3 graduation at the top and the 50 cm^3 graduation at the bottom of the burette?

2 Explain why it is not necessary to fill the burette to zero every time.

3 Outline safety precautions that you would suggest for an acid–alkali titration.

Method

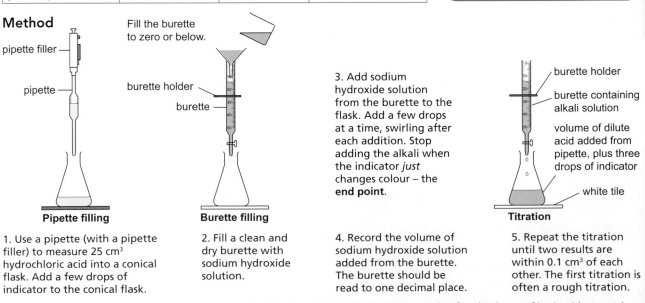

Fill the burette to zero or below.

pipette filler

pipette

burette holder

burette

3. Add sodium hydroxide solution from the burette to the flask. Add a few drops at a time, swirling after each addition. Stop adding the alkali when the indicator *just* changes colour – the **end point**.

burette holder

burette containing alkali solution

volume of dilute acid added from pipette, plus three drops of indicator

white tile

Pipette filling

Burette filling

Titration

1. Use a pipette (with a pipette filler) to measure 25 cm^3 hydrochloric acid into a conical flask. Add a few drops of indicator to the conical flask.

2. Fill a clean and dry burette with sodium hydroxide solution.

4. Record the volume of sodium hydroxide solution added from the burette. The burette should be read to one decimal place.

5. Repeat the titration until two results are within 0.1 cm^3 of each other. The first titration is often a rough titration.

FIGURE 2: Titration to measure the volume of sodium hydroxide solution that reacts with a fixed volume of hydrochloric acid.

Q ... acid-base indicators

Calculations (Higher tier)

Table 2 shows the results from the titration of 25.0 cm³ hydrochloric acid (unknown concentration) against 0.10 mol/dm³ sodium hydroxide solution.

TABLE 2: Example titration results.

Titration	rough	1	2	3	mean of 1, 2 and 3
Final burette reading (cm³)	18.1	17.0	32.6	25.4	
Initial burette reading (cm³)	2.0	1.5	17.0	10.0	
Volume of alkali used (cm³)	16.1	15.5	15.6	15.4	15.5

The results show that 25.0 cm³ of hydrochloric acid (unknown concentration) reacts completely with 15.5 cm³ of 0.10 mol/dm³ sodium hydroxide.

The equation

$$HCl(aq) + NaOH(aq) \rightarrow NaCl(aq) + H_2O(\ell)$$

shows that one mole of hydrochloric acid reacts with one mole of sodium hydroxide.

You can calculate the number of moles of sodium hydroxide used in the titration from its concentration (0.10 mol/dm³) and volume used (15.5 cm³). Note that you need to divide by 1000 to convert volume in cm³ to dm³.

moles of sodium hydroxide used = concentration (mol/dm³) × volume (dm³)

$$= 0.10 \times \frac{15.5}{1000} = 0.00155 \text{ mol}$$

This equals the number of moles of acid used. Therefore, there are 0.00155 mol of hydrochloric acid in 25 cm³.

The concentration of hydrochloric acid (mol/dm³) = moles ÷ volume (dm³)

$$= 0.00155 \div \frac{25}{1000}$$

$$= 0.062 \text{ mol/dm}^3$$

QUESTIONS

4 25.0 cm³ of 0.10 mol/dm³ sodium hydroxide neutralises 19.8 cm³ hydrochloric acid. Find the concentration of the acid.

5 10 cm³ of 0.2 mol/dm³ sodium hydroxide neutralises 16.0 cm³ sulfuric acid. What is the concentration of the acid?

Using equations

$$HCl(aq) + NaOH(aq) \rightarrow NaCl(aq) + H_2O(\ell)$$

shows that one mole of hydrochloric acid reacts with one mole of sodium hydroxide to produce one mole of sodium chloride and one mole of water.

Using the results in Table 2, what mass of sodium chloride is produced?

0.00155 mol hydrochloric acid reacted with 0.00155 mol of sodium hydroxide. They would have produced 0.00155 mol of sodium chloride and 0.00155 mol of water.

To calculate the mass of sodium chloride:

1. Calculate the relative formula mass of sodium chloride.

A_r values: Na = 23, Cl = 35.5

M_r NaCl = 23 + 35.5 = 58.5

2. Convert moles into mass.

Since M_r for NaCl is 58.5, 1 mol of sodium chloride has mass 58.5 g.

Therefore, 0.00155 mol of sodium chloride has mass 58.5 × 0.00155 = 0.091 g. This is the mass of sodium chloride produced.

QUESTIONS

6 What mass of potassium sulfate could be obtained from 25 cm³ of 0.5 mol/dm³ sulfuric acid? (Use A_r values: K = 39, S = 32, O = 16)

Q ... titration equations GCSE

Dynamic equilibrium

You will find out:

> what is happening when equilibrium is reached in a closed system

> how reaction conditions affect the amounts of reacting substances at equilibrium

> how changes in temperature and pressure affect the position of equilibrium

Fizzy or flat

The bubbles in cola are carbon dioxide. In an unopened bottle, most of the carbon dioxide is dissolved, but some is in the space above the drink. Open the bottle and bubbles appear in the drink as carbon dioxide escapes. Replace the top and the number of bubbles leaving the drink slowly decreases as the balance between the dissolved carbon dioxide and the gas above the drink is re-established.

FIGURE 1: The carbon dioxide dissolved in cola is in equilibrium with the carbon dioxide in the air space.

Dynamic equilibrium

If a reaction is reversible, it will go both ways. Some products will turn back into reactants.

When a reaction reaches equilibrium, the rate of the forward reaction equals the rate of the backward reaction. The concentrations of reactant and product are constant.

Example

Hydrogen and iodine are mixed, in a sealed container – this is a **closed system**. A reversible reaction, one that can go in both directions, happens.

Forward reaction: $H_2(g) + I_2(g) \rightarrow 2HI(g)$
Backward reaction: $H_2(g) + I_2(g) \leftarrow 2HI(g)$

The two reactions occur simultaneously, shown by the equation:

$H_2(g) + I_2(g) \rightleftharpoons 2HI(g)$

> H_2 and I_2 molecules collide to produce HI molecules. Initially the reaction is fast. As the reaction proceeds, H_2 and I_2 concentrations decrease and the reaction slows.

> At the same time, HI molecules break down into H_2 and I_2. Initially the reaction is slow, but becomes faster.

> Eventually, the rate of forward reaction equals the rate of the backward reaction. The concentrations of H_2, I_2 and HI do not change.

> The reaction does not stop. HI molecules continue to form and to break down, but at the same rate so the concentrations do not change.

The system is in **dynamic equilibrium** and the **position of equilibrium** has been reached.

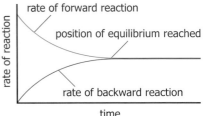

FIGURE 2: At equilibrium the rate of the forward reaction equals the rate of the backward reaction.

QUESTIONS

1 Suggest why combustion is not a reversible reaction.

2 'The chemical reaction is in a position of dynamic equilibrium'. Describe what this statement means.

FIGURE 3: The runner and the treadmill belt are moving at the same speed, but in opposite directions. They are in dynamic equilibrium.

Q ... chemical equilibria

Changing the position of equilibrium

Counteracting change

If a system is in dynamic equilibrium and the conditions are changed, the position of the equilibrium moves to counteract the change. This is **Le Chatelier's principle**.

Changing temperature

> In reactions that are **endothermic** – increasing temperature increases yield; decreasing temperature decreases the yield.

> In reactions that are **exothermic** – increasing temperature decreases yield; decreasing temperature increases the yield.

Consider the reaction of ethene with steam:

promoted by low temperatures
forward reaction is exothermic (heats surroundings)

$$C_2H_4(g) + H_2O(g) \rightleftharpoons C_2H_5OH(g)$$

backward reaction is endothermic (heated by surroundings)
promoted by high temperatures

Raising the temperature increases the proportion of ethene and steam in the equilibrium mixture. Lowering the temperature increases the proportion of ethanol in the equilibrium mixture.

Catalysts

A catalyst does not change the position of equilibrium. However, a catalyst does enable the position of equilibrium to be reached more quickly.

Many chemicals are made in reversible reactions in chemical plants (chemical factories). Temperature and pressure are controlled to give the best yield of their product at equilibrium. A catalyst is used to achieve this more quickly.

◉ QUESTIONS

3 Nitrogen monoxide is produced in car engines in the endothermic reaction: $N_2(g) + O_2(g) \rightleftharpoons 2NO(g)$. Describe how temperature change affects the position of equilibrium.

4 Explain why a catalyst does not alter the position of a dynamic equilibrium.

5 Suggest why catalysts are so important in industrial processes.

Changing pressure

Increasing the pressure of a gaseous reaction will favour the reaction that produces the least number of molecules as shown by the symbol equation for that reaction.

For example, the hydration of ethene to make ethanol gives one mole of products from two moles of reactants.

promoted by higher pressures
forward reaction produces fewer molecules

$$C_2H_4(g) + H_2O(g) \rightleftharpoons C_2H_5OH(g)$$

backward reaction produces greater number of molecules
promoted by lower pressures

Raising the pressure increases the proportion of ethanol in the equilibrium mixture. Lowering the pressure increases the proportion of ethene and steam in the equilibrium mixture.

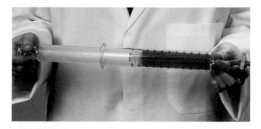

FIGURE 4: Mixtures of brown NO_2 and colourless N_2O_4 in equilibrium at different pressures.

◉ QUESTIONS

6 A syringe was filled with nitrogen dioxide (see Figure 4). It was left until the reaction $2NO_2(g) \rightleftharpoons N_2O_4(g)$ reached equilibrium. Pushing in the plunger made the gas look darker brown. After a short while, the reaction mixture lightened in colour. Explain the observations.

Making ammonia

You will find out:
> how ammonia is made in the Haber process
> how the conditions of the reaction are controlled to give the maximum yield

Guano fertiliser

In the 19th century, transporting ship loads of bird droppings, called guano, was a profitable business – if you could stand the smell. It contained nitrogen compounds and was used as a fertiliser to increase crop yields. Nowadays, most farmers use manufactured nitrogen compounds as fertilisers.

FIGURE 1: Fertiliser from cormorant guano is still produced commercially in Peru.

The Haber process

Nitrogen fertilisers

Plants need nitrogen to grow healthily. Symptoms of nitrogen deficiency include leaves and stems becoming light green or yellow and plant growth slowing.

Making ammonia

Ammonia is manufactured in the **Haber process**. It is made by the reversible reaction of nitrogen with hydrogen:

$$N_2(g) + 3H_2(g) \rightleftharpoons 2NH_3(g)$$

Figure 3 shows the flow diagram for the Haber process. The nitrogen is obtained from air by fractional distillation. The hydrogen is usually produced from methane in natural gas.

Nitrogen, hydrogen and ammonia are in a state of dynamic equilibrium. The relative amounts present depend on the temperature and pressure of the reaction mixture.

> Under the conditions used, the position of the equilibrium is towards the reactants. Only 15% of reactants are converted to ammonia.

> The equilibrium mixture, containing hydrogen, nitrogen and ammonia, is cooled until the ammonia liquefies and can be separated.

> Unreacted nitrogen and hydrogen are recycled and added to the reaction mixture.

FIGURE 2: The maize on the left has been grown in soil with a higher nitrogen content than the maize on the right.

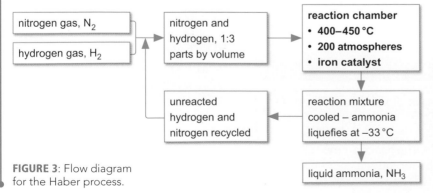

FIGURE 3: Flow diagram for the Haber process.

nitrogen gas, N_2

hydrogen gas, H_2

→ nitrogen and hydrogen, 1:3 parts by volume

→ **reaction chamber**
- **400–450 °C**
- **200 atmospheres**
- **iron catalyst**

unreacted hydrogen and nitrogen recycled

reaction mixture cooled – ammonia liquefies at −33 °C

liquid ammonia, NH_3

QUESTIONS

1 Describe how you would recognise nitrogen deficiency in plants.

2 Explain why ammonia can be separated from hydrogen and nitrogen by cooling to −50 °C.

The yield (Higher tier)

The Haber process uses conditions that give the highest cost-effective yield.

> The equilibrium position of the reaction

$$N_2(g) + 3H_2(g) \rightleftharpoons 2NH_3(g)$$

must be pushed as far to the right as possible to maximise yield.

> The reaction must proceed as quickly as possible.

> Energy requirements must be as low as possible.

TABLE 1: Conditions used in the Haber process.

Condition	Effect on the position of equilibrium for $N_2(g) + 3H_2(g) \rightleftharpoons 2NH_3(g)$	Effect on the rate of reaction
temperature	The forward reaction ($\longrightarrow$) is exothermic. The backward reaction ($\longleftarrow$) is endothermic. Increasing the temperature pushes the reaction in the endothermic direction. A low temperature is needed to increase the yield of ammonia.	High temperature increases the rate of reaction. The reaction mixture reaches equilibrium quicker. Low temperature decreases the rate of reaction and it takes longer to reach equilibrium.
pressure	There are four moles of molecules on the left-hand side of the equation and only two on the right. Therefore, increasing the pressure favours the forward reaction and increases the yield of ammonia.	Increased pressure pushes the molecules closer together, leading to more frequent successful collisions and faster reaction. The equilibrium is reached quicker.
catalyst	A catalyst has no effect on the position of equilibrium.	Using a suitable catalyst increases the rate of reaction and equilibrium is reached quicker.

Conditions for the process

The optimum yield is the maximum amount of ammonia that can be produced for the minimum use of energy. This is achieved by:

> a compromise temperature of 450 °C. Equilibrium is reached quickly, although the yield is lower.

> a compromise pressure of 200 atmospheres. Higher pressures would give higher yields and faster, but need expensive equipment and maintenance.

> a catalyst of finely divided iron.

> **QUESTIONS**

3 Explain why a temperature of 450 °C is used in the Haber process.

4 Describe the energy-saving measures used in the Haber process.

Making the hydrogen

Hydrogen for the Haber process is made from methane (natural gas) in a process called **steam reforming**. The reaction is in two steps.

Step one: $CH_4(g) + H_2O(g) \rightleftharpoons CO(g) + 3H_2(g)$

The reaction to make hydrogen is endothermic and carried out at 700–1000 °C with a nickel catalyst.

Step two: $CO(g) + H_2O(g) \rightleftharpoons CO_2(g) + H_2(g)$

The reaction to make hydrogen is exothermic and carried out at a compromise temperature of 250 °C with a metal catalyst.

> **QUESTIONS**

5 In the two steps for making hydrogen, described on the left, two different temperatures are used. Explain why these temperatures are chosen.

Q ... Haber process yield GCSE ... steam reforming methane

Preparing for assessment: Analysing and interpreting data

To achieve a good grade in science, you not only have to know and understand scientific ideas, but you need to be able to apply them to other situations and investigations. These tasks will support you in developing these skills.

✹ Cashing in on ammonia

Making fertilisers is big business and essential to feed the world's increasing population. Ammonia, NH_3, is a key ingredient of fertilisers, but ammonia is notoriously difficult to make.

The reaction to make ammonia from hydrogen and nitrogen is reversible. To make most profit, manufacturers use a temperature of 450 °C, a pressure of 200 atmospheres and an iron catalyst.

Some students were set a project to find out why these conditions are used.

Hypothesis

The students' hypothesis is that a temperature of 450 °C, a pressure of 200 atmospheres and an iron catalyst give the highest yield of ammonia possible.

Method

They carried out an online search for secondary data and found information showing how temperature and pressure affect the yield of ammonia.

20 Million Acres of Proven Performance

Hasta Gro Plant

6-12-6 LIQUID PLANT FOOD PLUS
Contains Plant Food Supplements PLUS Humate
Humic Acid and Medina Soil Activator

*Incluye instrucc
en español*

GUARANTEED ANALYSIS
Total Nitrogen (N) 6.00%
Available Phosphoric Acid (P₂O₅) 12.00%
Soluble Potash (K₂O) 6.00%
Copper (Cu) .... .02%
    .02% Chelated Copper
Iron (Fe) .... .05%
    .05% Chelated Iron
Manganese (Mn) .... .05%
    .05% Chelated Manganese
Molybdenum (Mo) .... .0005%
Zinc (Zn) .... .05%
    .05% Chelated Zinc
Humic Acid 1.00%

Derived from Urea, Phosphoric Acid, Potassium Hydroxide,
Aqua Ammonia, EDTA Copper, EDTA Iron, EDTA Manganese,
Sodium Molybdate, EDTA Zinc and Leonardite Ore.

ONE QUART (.95L) NET WEIGHT 2.5 LBS. (1.13Kg)
Un cuarto de galón (946ml) Peso Neto 2.5 libras (1.13Kg)

✹ Collecting secondary data

Industrial scientists mixed nitrogen and hydrogen gas in a 1:3 ratio and kept the mixture at three different temperatures. They changed the pressure to different values and measured the percentage of ammonia at each pressure.

These are some of the scientists' results:

Pressure (atmospheres)	Percentage of ammonia at equilibrium (by volume)		
	100 °C	300 °C	500 °C
10	–	15	1
25	92	27	3
50	94	39	6
100	97	52	11
200	98	67	18
400	99	80	32
1000	–	93	57

No readings were available for the 100 °C mixtures at 10 and 1000 atmospheres.

1. Draw a graph of pressure against percentage of ammonia at 100 °C. Plot, on the same axes, two more best fit lines for the results at 300 °C and 500 °C.

> Make sure that you can identify the independent variable, the dependent variable and the controlled variables. You will need these to plot your graph.

> The independent variable goes on the *x*-axis. The dependent variable goes on the *y*-axis. Plot the results for each temperature separately and draw a line of best fit. Label each line.

✸ Analysing data

2. Identify any anomalous results.

3. What is the relationship between the pressure (independent variable) and the percentage of ammonia produced (dependent variable)?

4. What is the relationship between the temperature (independent variable) and the percentage of ammonia produced?

5. How far do the data support the hypothesis?

6. Write a conclusion from these results.

7. If your conclusion does not support the hypothesis and the conditions used to make ammonia commercially, can you think of reasons why? Suggest what further information might be needed to explain why the conditions in the hypothesis are used.

> Anomalous results are results that do not fit the general pattern. This measurement would need to be repeated.

> Is there enough evidence to arrive at a firm conclusion?

> This should link to the hypothesis. Make sure that you justify your conclusion by stating simply what the evidence shows.

✸ Using scientific explanations

Ammonia is made in the reaction $N_2(g) + 3H_2(g) \rightleftharpoons 2NH_3(g)$

Reacting nitrogen and hydrogen to make ammonia is an exothermic reaction.

You can assume that the scientists' results are valid.

8. Ammonia was constantly being broken back down to hydrogen and nitrogen when the percentage of ammonia was measured. Explain why this did not affect the results.

9. Changing the temperature affects the yield of ammonia. Explain why.

10. Explain how and why changes in pressure affect the yield of ammonia.

11. Using a catalyst decreases the time taken for the reaction to reach equilibrium. Suggest why this is good for the manufacturer.

> Your explanation needs to consider why the position of the equilibrium changes when the temperature changes.

> Again, your answer needs to consider how the position of the equilibrium changes and why.

> Think about economic considerations.

✸ Connections

How Science Works

- Collect primary and secondary data
- Select and process primary and secondary data
- Analyse and interpret primary and secondary data
- Use scientific models and evidence to develop hypotheses, arguments and explanations

Science ideas

C3.5 The production of ammonia

Alcohols

Slow signals

The alcohol in everyday life is the compound, ethanol. In small amounts, ethanol can help people relax socially, but it is also a depressant. It slows down signals in the nervous system by affecting their passage across the synapses in the brain. This is why it is dangerous to drink and drive.

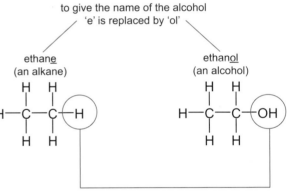

Don't drink and drive

You will find out:
> about alcohols
> about reactions and uses of alcohols
> how vinegar is made from alcohol

FIGURE 1: Drivers break the law in the UK if they have more than 80 mg ethanol in 100 cm³ of their blood.

Names, formulae and structures

The alcohols are a family of organic chemicals forming a **homologous** series. They contain the **hydroxyl** group (–OH). This is known as a **functional group** because it gives alcohols their characteristic properties.

Alcohols have the general formula $C_nH_{2n+1}OH$.

The first part of an alcohol's name is the name of the alkane, but without the last 'e'. The end is always –ol, showing that it contains the hydroxyl functional group.

Figure 2 shows an example of the difference between an alkane and an alcohol. A hydrogen atom in the alkane has been replaced by a hydroxyl group.

Bonding sequence

Table 1 shows the sequence in which the atoms are bonded together. The semi-structural formula is written carbon atom by carbon atom.

to give the name of the alcohol 'e' is replaced by 'ol'

ethane (an alkane)   ethanol (an alcohol)

hydrogen atom is replaced by a hydroxyl group

FIGURE 2: Displayed structures for the alkane, ethane and the alcohol, ethanol.

TABLE 1: The first three members of the alcohol family:

Name	Molecular formula	Semi-structural formula	Displayed structure
methanol	CH_3OH	CH_3OH	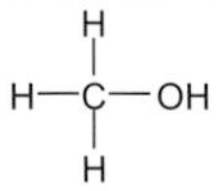
ethanol	C_2H_5OH	CH_3CH_2OH	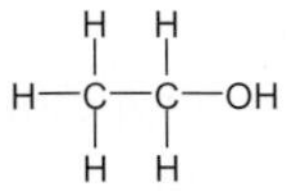
propanol	C_3H_7OH	$CH_3CH_2CH_2OH$	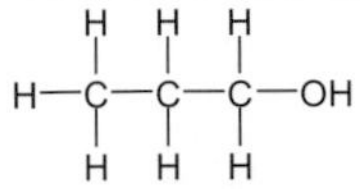

QUESTIONS

1 Add another row to Table 1, for the alcohol, butanol. Suggest what should be in the three blank boxes.

2 Describe the difference in structure between butane and butanol.

Q ... introduction alcohols

Reactions and uses of alcohols

Solubility in water

Methanol, ethanol and propanol dissolve in water to give neutral solutions (pH = 7). This is unlike hydroxides such as sodium hydroxide, NaOH, which dissolve in water to give alkaline solutions (pH > 7). It is the covalent bonding in alcohol molecules that gives this solubility.

$NaOH(s) + water \rightarrow Na^+(aq) + OH^-(aq)$ pH > 7 (hydroxide ions in solution)

$C_2H_5OH(l) + water \rightarrow C_2H_5OH(aq)$ pH = 7 (no hydroxide ions in solution)

Reactions with sodium

Like water, methanol, ethanol and propanol react with sodium metal to produce hydrogen. For example,

sodium + ethanol → sodium ethoxide + hydrogen

$2Na(s) + 2C_2H_5OH(l) \rightarrow 2C_2H_5ONa(aq) + H_2(g)$

Reactions with oxygen

Methanol, ethanol and propanol burn readily in oxygen to make carbon dioxide and water.

$2CH_3OH(l) + 3O_2(g) \rightarrow 2CO_2(g) + 2H_2O(g)$

$C_2H_5OH(l) + 3O_2(g) \rightarrow 2CO_2(g) + 3H_2O(g)$

$2C_3H_7OH(l) + 11O_2(g) \rightarrow 6CO_2(g) + 8H_2O(g)$

The combustion reactions are highly exothermic. This makes alcohols useful as fuels.

TABLE 2: Energy given out when one mole of fuel burns in oxygen.

Fuel	methane	ethane	propane	methanol	ethanol	propanol
Energy released (kJ)	890	1560	2170	726	1300	2020

QUESTIONS

3 Why do different alcohols have similar chemical properties?

4 Write word and symbol equations for the combustion of butanol, C_4H_9OH.

5 Describe the trends in Table 2.

Uses

The main uses of alcohols are as fuels, in drinks and as solvents.

> Ethanol is commonly used as motor fuel, often as a biofuel additive to petrol. Ordinary cars in the UK can run on 5% blends of ethanol. Most cars in Brazil and the US run on 10% ethanol blends.

> Ethanol is the main alcohol in alcoholic drinks.

> Alcohols are useful solvents for some substances that are water insoluble. Ethanol is used as a solvent in paints, medicines and perfumes. Propanol is used as a solvent in cosmetics, medicines and cleaning fluids.

FIGURE 3: Ethanol is highly flammable. What precautions must you take when using ethanol?

Wine into vinegar

Vinegar is a mixture of ethanoic acid and water, plus flavourings. It is made from cheap wine or cider.

> Ethanol in wine, for example, is oxidised to ethanoic acid by microbial action. A bacterium called *Acetobacter* catalyses the reaction of ethanoic acid with oxygen in the air.

$C_2H_5OH + O_2 \rightarrow CH_3COOH + H_2O$

$$H-\underset{\underset{H}{|}}{\overset{\overset{H}{|}}{C}}-\underset{\underset{H}{|}}{\overset{\overset{H}{|}}{C}}-OH + O_2 \rightarrow H-\underset{\underset{H}{|}}{\overset{\overset{H}{|}}{C}}-\underset{\underset{OH}{|}}{C}=O + H_2O$$

> Ethanol can also be oxidised to ethanoic acid by chemical oxidising agents, such as acidified potassium dichromate(VI) solution.

QUESTIONS

6 Make molecular models to represent the oxidation of ethanol to (a) carbon dioxide and water (b) ethanoic acid.

Carboxylic acids

You will find out:

> about carboxylic acids and esters

> the properties of carboxylic acids

> about weak and strong acids

Savoury and sweet

The distinctive taste of the UK's favourite pickle comes from a dilute solution of ethanoic acid – a carboxylic acid otherwise known as vinegar. The same carboxylic acid is used to make fruit flavours that are added to sweets. If the ingredients include 'natural flavourings', the flavours were made by nature. 'Flavourings' suggest that they were made in a laboratory.

FIGURE 1: Esters give the flavours. How were the esters made?

Names, formulae and structures

The carboxylic acids are a homologous series of organic acids. They all contain the functional group –COOH and have the general formula $C_nH_{2n+1}COOH$.

The first part of a carboxylic acid's name is the name of the alkane, but without the last 'e'. The end is always –oic. This shows that it contains the carboxylic acid functional group.

Figure 2 shows an example of the difference between an alkane and a carboxylic acid. A CH_3 group in the alkane has been replaced by a COOH group.

to give the name of the carboxylic acid 'e' is replaced by 'oic'

ethane (an alkane) → ethanoic acid (a carboxylic acid)

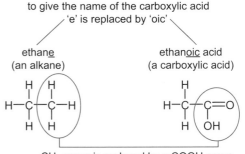

CH_3 group is replaced by a COOH group

FIGURE 2: Displayed structures for the alkane, ethane and the carboxylic acid, ethanoic acid.

TABLE 1: The first three members of the carboxylic acid family.

Name	Molecular formula	Semi-structural formula	Displayed structure
methanoic acid	HCOOH	HCOOH	H—C—OH ‖ O
ethanoic acid	CH_3COOH	CH_3COOH	H H—C—C—OH H O
propanoic acid	C_2H_5COOH	CH_3CH_2COOH	H H H—C—C—C—OH H H O

QUESTIONS

1 Describe the differences in the structures of carboxylic acids and alcohols.

2 Add another row to Table 1, for butanoic acid. Suggest what should be in the three blank boxes.

Reactions

Solubility in water

Methanoic, ethanoic and propanoic acids all dissolve in water to produce acidic solutions.

Reactions with carbonates

Like all acids, solutions of carboxylic acids in water react with carbonates to produce carbon dioxide gas. For example, when ethanoic acid is added to sodium carbonate:

ethanoic acid + sodium carbonate → sodium ethanoate + water + carbon dioxide

$2CH_3COOH + Na_2CO_3 \rightarrow 2CH_3COONa + H_2O + CO_2$

Q … carboxylic acid

Reactions with alcohols

Carboxylic acids and alcohols react together to make esters. A few drops of sulfuric acid are needed as a catalyst. The general reaction is:

alcohol + carboxylic acid → ester + water

For example, ethanol reacts with ethanoic acid to produce ethyl ethanoate and water.

$C_2H_5OH(\ell) + CH_3COOH(aq) \rightarrow CH_3COOC_2H_5 (aq) + H_2O(\ell)$

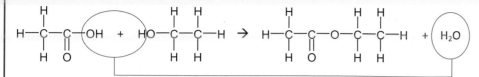

Different carboxylic acids and alcohols make different esters, but all esters contain the functional group –COO–. They are volatile (evaporate easily at room temperature) and have distinctive smells such as banana and pineapple. The name of an ester tells you the alcohol and carboxylic acid used.

Esters are used as flavourings and in perfumes.

TABLE 2: Reactants needed to make some esters.

Carboxylic acid	Alcohol	Ester	Smell
ethanoic acid, CH_3COOH	pentanol, $C_5H_{11}OH$	pentyl ethanoate	pears
ethanoic acid, CH_3COOH	octanol, $C_8H_{17}OH$	octyl ethanoate	bananas
butanoic acid, C_3H_7COOH	pentanol, $C_5H_{11}OH$	pentyl butanoate	strawberries

⦿ QUESTIONS

3 Suggest some tests to distinguish between (a) an alcohol and a carboxylic acid (b) an ester and a carboxylic acid.

4 Draw the displayed structure for the ester that smells of (a) pears (b) strawberries.

Strong and weak acids (Higher tier)

Acids dissolve in water to give a solution containing hydrogen ions, H^+.

Hydrochloric acid and sulfuric acid ionise completely in water. For example:

$HCl(g) + water \rightarrow H^+(aq) + Cl^-(aq)$

Acids which ionise completely in water are called strong acids. Not all acids are strong.

Carboxylic acids are weak acids because they do not ionise completely in water. A solution of ethanoic acid in water contains ethanoic acid molecules, ethanoate ions and hydrogen ions.

$CH_3COOH(\ell) + water \rightarrow CH_3COOH(aq)$

Then a dynamic equilibrium is established between un-ionised molecules, $CH_3COOH(aq)$, and ions, $CH_3COO^-(aq)$ and $H^+(aq)$:

$CH_3COOH(aq) \rightleftharpoons CH_3COO^-(aq) + H^+(aq)$

The position of the equilibrium lies to the left of the equation and there are relatively fewer hydrogen ions in the solution.

Aqueous solutions of weak acids have a higher pH value than aqueous solutions of strong acids with the same concentration. For example:

> 0.1 mol/dm³ solution of hydrochloric acid has a pH of 1

> 0.1 mol/dm³ solution of ethanoic acid has a pH of about 3

⦿ QUESTIONS

5 Describe, in terms of the particles present, what is happening in an aqueous solution of ethanoic acid.

strong acid
0.1 mol/dm³
hydrochloric acid

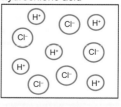

weak acid
0.1 mol/dm³
ethanoic acid

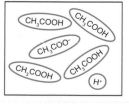

FIGURE 3: Examples of strong and weak acids.

Preparing for assessment: Applying your knowledge

To achieve a good grade in science, you not only have to know and understand scientific ideas, but you need to be able to apply them to other situations and investigations. These tasks will support you in developing these skills.

✺ Latte or americano, regular or decaf?

Mark and Tim are opting for a quick cup of their favourite coffee before the afternoon GCSE Chemistry exam starts. Mark has ordered a regular latte and Tim has ordered a decaf americano. Tim knows that ordinary coffee makes him jittery and his doctor has suggested he only drinks decaffeinated coffee. 'Decaf' is short for decaffeinated.

All coffee beans contain caffeine. It is a natural pesticide in the coffee plant that paralyses and kills certain insects feeding on the plant. The caffeine in coffee acts as a stimulant in humans. It affects the central nervous system. Mark's latte will, he hopes, keep him alert throughout his chemistry exam.

Caffeine has an adverse effect on some people and for Tim, decaffeinated coffee is the best choice.

Decaffeinated coffee is made from normal coffee beans in food factories. The beans are usually steamed or soaked to swell them and then a solvent is used to dissolve out the caffeine.

The manufacture of many food products involves the use of solvents to dissolve, extract or purify the product.

Caffeine is soluble in several organic liquids, but manufacturers must use a non-hazardous solvent that does not change the taste of the coffee. Ethyl ethanoate is commonly used. Ethyl ethanoate occurs naturally in many fruits and vegetables. It makes up part of the smell of many fruits. Because of this, decaf coffee producers are able to label their product as 'naturally decaffeinated'.

✳ Task 1

Mark and Tim have made ethyl ethanoate in their chemistry lessons. The ethyl ethanoate used to extract caffeine from coffee beans is made in chemical plants (factories) using the same reaction.

(a) To which family of chemicals does ethyl ethanoate belong?

(b) Apart from producing decaf coffee, how is this family of chemicals mostly used?

(c) Which two other families of chemicals are used to make chemicals such as ethyl ethanoate?

✳ Task 2

They have been revising the chemistry involved in making ethyl ethanoate for their GCSE exam.

(a) Name the two reactants they would have used, in class, to make ethyl ethanoate and draw their displayed formulae.

(b) Draw the displayed formula for ethyl ethanoate.

(c) Why did Mark and Tim add a few drops of sulfuric acid to their reactants?

(d) How were they able to detect the ethyl ethanoate in their experiment?

✳ Task 3

Mark and Tim have read on a website that "The reaction to make ethyl ethanoate is a reversible reaction. When the reaction reaches equilibrium, the yield of ethyl ethanoate is 65% at room temperature."

(a) Explain what 'reversible reaction' means.

(b) What is happening to the reaction when equilibrium is reached?

(c) Why is this reversible reaction a nuisance to ethyl ethanoate manufacturers?

(d) Why was it important for the website to give the temperature, when quoting the yield?

✳ Maximise your grade

	Answer includes showing that you...
	recall that esters are used as fruit flavourings and smells.
	recall the chemical formulae of the first three alcohols and carboxylic acids.
	can describe what happens in a reversible reaction.
E	can explain that reactants and products are present at equilibrium.
	can describe how temperature affects yield at equilibrium.
C	recognise alcohols and carboxylic acids from their names.
	can draw the displayed formulae for three alcohols and carboxylic acids.
	can describe the reaction to produce ethyl ethanoate.
A	can draw the displayed formula for ethyl ethanoate.
	can explain the rates of forward and backward reactions at equilibrium.

Checklist C3.4–3.6

To achieve your forecast grade in the exam you will need to revise

Use this checklist to see what you can do now. Refer back to the relevant topics in this book if you are not sure. Look across the three columns to see how you can progress. Bold text means Higher tier only.

Remember that you will need to be able to use these ideas in various ways, such as:

> interpreting pictures, diagrams and graphs

> applying ideas to new situations

> explaining ethical implications

> suggesting some benefits and risks to society

> drawing conclusions from evidence you are given.

Look at pages 188–209 for more information about exams and how you will be assessed.

To aim for a grade E	To aim for a grade C	To aim for a grade A
Use a flame test to identify some metal ions.	Identify some metal ions from their reactions with sodium hydroxide solution.	Write ionic equations for the precipitation of insoluble hydroxides.
Know that carbonates give off carbon dioxide gas with dilute acid.	Identify carbonate, sulfate and halide ions from chemical tests.	Write balanced symbol equations and ionic equations for identification reactions.
Use instructions or help to carry out an acid–alkali titration safely.	Know how to carry out an acid–alkali titration and record the results.	**Calculate chemical quantities in a titration using concentrations and masses.**
Know that some reactions are reversible and can go both ways.	Understand that, when an equilibrium is reached in a closed system, both reactants and products are present.	Know that, at equilibrium, the forward and backward reactions occur at the same rate.
Name examples of transition metals and some of their properties.	Interpret transition metal data and describe their properties.	
Know the factors that affect a chemical equilibrium.	Understand how the amounts of all reactants at equilibrium depend on temperature and pressure (for gaseous reactions).	
Know that ammonia is a gas made from hydrogen and nitrogen.	Know how the raw materials for the Haber process are obtained.	

To aim for a grade E   To aim for a grade C   To aim for a grade A

Know that ammonia is made in a chemical plant (factory).

Know the conditions used in the Haber process.

Know that the reaction to make ammonia is reversible.

Understand that these conditions are chosen to give the highest yield.

Describe and evaluate how changing the temperature and pressure affect the yield of ammonia in the Haber process.

Know that an iron catalyst in used in the Haber process.

Understand that the catalyst decreases the time taken to reach equilibrium but does not change the position of the equilibrium.

Know the names of the first three alcohols and recognise their functional group.

Know the chemical and displayed formulae for the first three alcohols.

Know how to handle alcohols safely.

Know that alcohols are used as fuels and solvents.

Write word equations for the combustion reactions of alcohols.

Know how alcohols react with sodium and oxidising agents (or microbes).

Write balanced symbol equations for the combustion reactions of the first three alcohols.

Know that ethanoic acid is a weak acid found in vinegar.

Recognise the functional group for ethanoic acid.

Understand that ethanoic acid is a carboxylic acid.

Name and give the chemical formula of the first three carboxylic acids.

Know that ethanoic acid has typical acid properties.

Understand that carboxylic acids dissolve in water and react with carbonates and alcohols.

Understand that weak acids do not ionise completely when dissolved in water.

Understand that solutions of weak acids have higher pH values than strong acids with the same concentration.

Know that esters have distinctive fruity smells.

Draw structural formulae to show how ethanol and ethanoic acid react to produce the ester ethyl ethanoate.

Recognise the functional group for esters.

Know how esters are used.

In the examination, data and the periodic table will be given on a separate sheet.
You will be expected to select appropriate data from the sheet.

1. Students carried out a titration to make sodium chloride. The chemical reaction is:

$NaOH(aq) + HCl(aq) \rightarrow NaCl(aq) + H_2O(\ell)$

AO1 **(a)** The diagram below shows the apparatus that they used.

Name A and B. [2]

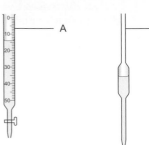

AO2 **(b)** The students measured 25 cm³ of dilute hydrochloric acid into a conical flask and added a few drops of universal indicator. What is the colour of the indicator in acid? [1]

AO2 **(c)** They slowly added sodium hydroxide solution until the mixture was neutral. Explain how they knew that it is neutral. [1]

AO2 **(d)** What test can they do, to show that the product contains sodium ions? Give the test and the positive result. [2]

AO2 **(e)** What test can they do to show that the product contains chloride ions? Give the test and the result. [2]

2. Some false nails are made from acrylic. Nail polish remover for acrylic nails contains ethyl ethanoate.

AO1 **(a)** What type of organic compound is ethyl ethanoate? [1]

AO2 **(b)** Ethyl ethanoate is produced using ethanol and ethanoic acid. Give the missing information for A, B, C and D in the table below. [4]

Substance	Homologous series	Molecular formula	Displayed formula
ethanol	A	C_2H_5OH	B
ethanoic acid	carboxylic acid	C	D

3. Most permanent hair dyes contain ammonium compounds to make the dye bond to the hair. Ammonia is made in chemical plants using the Haber process.

AO1 **(a) (i)** Name the two elements that are used to make ammonia in the Haber process. [2]

(ii) Where do these elements come from? [2]

(iii) Give the molecular formula of ammonia. [1]

AO1 **(b)** The chemical reaction to make ammonia is reversible. What does it mean if a reaction is reversible? [1]

AO1 **(c)** What temperature, pressure and catalyst are used in the Haber process? [3]

AO3 **(d)** Use information from the table, below, to explain how ammonia can be removed from the unreacted hydrogen and nitrogen. [2]

Substance	Boiling point (°C)
hydrogen	-253
nitrogen	-196
ammonia	-33

4. Many people use salt substitutes instead of normal table salt for health reasons.

A chemist analysed two salt substitutes, as well as normal table salt. She used solutions. In Test 3, the chemist added a few drops of nitric acid, followed by a few drops of silver nitrate solution to each sample. Her results are shown in the table.

Salt	Flame test	Test 3
salt substitute A	yellow, with lilac	white precipitate
salt substitute B	lilac and red	white precipitate
table salt	yellow	white precipitate

AO3 **(a)** Name the metal ions in

(i) salt substitute A

(ii) salt substitute B

(iii) table salt. [5]

AO3 **(b)** Which non-metal ion did the chemist detect? [1]

AO1 recall the science   AO2 apply your knowledge   AO3 evaluate and analyse the evidence

✳ WORKED EXAMPLE – Foundation tier

A white powder has been recovered from a crime scene. Forensic officers do not know if it is harmless, such as chalk, or a banned substance.

These are the results of some initial lab tests carried out on the powder.

Testing for:	Test	Result
metal ion	flame test	red
	adding sodium hydroxide solution	white precipitate forms
		– did not dissolve when more sodium hydroxide solution was added
non-metal ion	carbonate test	no gas was given off so limewater test not used
	halide test (chloride, bromide and iodide)	no change
	sulfate test	white precipitate formed

(a) (i) Which metal ion is present in the white powder? [1]

calcium

(ii) What is the evidence for your conclusion (you need to describe the positive test)? [2]

Calcium gives a red flame test.

(b) (i) Which non-metal ion is present in the white powder? [1]

Sulfate ions

(ii) What is the evidence for this conclusion (you need to describe the positive test)? [1]

Sulfates make a white precipitate with barium chloride. Some hydrochloric acid needs to be added first.

(c) Some students suggested that the white powder could be a copper salt. Give one reason why the white powder is not a copper salt, other than the test results in the table. [2]

Copper is a transition metal and transition metals have coloured compounds. Copper salts are blue salts like copper sulfate and this powder is white.

Calcium is correct, but it would be more accurate to say calcium ions.

This is correct and gains one mark. Again, it would be better to say calcium ions.

There are two marks for this question, so look for two points in your answer. Calcium ions also produce a white precipitate with sodium hydroxide solution that does not dissolve in excess sodium hydroxide solution.

This is correct and receives the mark.

The candidate has named correctly the chemicals used in the test. However, the candidate needs to be accurate when describing them. The barium chloride has to be dissolved in water to make the test work, so the answer should refer to barium chloride solution. The hydrochloric acid must be dilute, so the answer should refer to dilute hydrochloric acid. Adding concentrated hydrochloric acid will be hazardous.

This is a good answer and gains both marks.

In the examination, data and the periodic table will be given on a separate sheet.
You will be expected to select appropriate data from the sheet.

1. An accident at a soap-making factory has resulted in too much potassium hydroxide being added to a batch of soft soap.

The factory's chemist titrated the contaminated soap solution against 0.1 mol/dm³ hydrochloric acid. He used 25 cm³ samples of acid. The results are shown in the table.

Titration	rough	trial 1	trial 2	trial 3
Final burette reading (cm³)	18.50	34.60	17.80	33.95
Initial burette reading (cm³)	1.00	18.50	1.60	17.80
Volume alkali used (cm³)	17.50	16.10	16.20	16.15

AO1 **(a)** Suggest a suitable indicator. [1]

AO2 **(b)** Calculate the mean volume of contaminated soap solution used to neutralise 25 cm³ 0.1 mol/dm³ hydrochloric acid. [1]

AO2 **(c)** Calculate how many moles are in 25 cm³ 0.1 mol/dm³ hydrochloric acid. [2]

AO2 **(d)** The equation for the neutralisation reaction is
KOH(aq) + HCl(aq) → KCl(aq) + H₂O(ℓ)

How many moles of potassium hydroxide react with 25 cm³ 0.1 mol/dm³ hydrochloric acid? [1]

AO2 **(e)** What is the concentration of the potassium hydroxide in the contaminated soap solution? Give your answer in mol/dm³. [2]

2. Flavourings are common in confectionery. Fruit flavours can be natural or made in a laboratory. They are made from chemicals called esters.

(a) Esters are made when two different types of chemicals react together.

AO1 **(i)** Name the two types of chemicals. [2]

AO2 **(ii)** Name the ester made when ethanol reacts with ethanoic acid and draw its displayed formula. [2]

AO1 **(iii)** Which group of atoms are found in all esters? [1]

AO2 **(b)** Ethanoic acid has similar properties to other acids. How would you expect it to react with:

(i) universal indicator
(ii) an alkali
(iii) a carbonate? [3]

3. Ammonia is made in the Haber process. The reaction is:

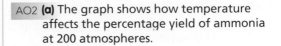

$$N_2(g) + 3H_2(g) \rightarrow 2NH_3(g)$$

The forward reaction is exothermic.

AO2 **(a)** The graph shows how temperature affects the percentage yield of ammonia at 200 atmospheres.

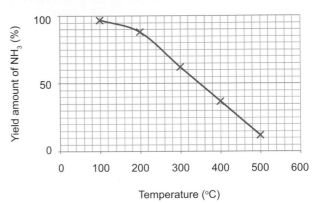

(i) What is the percentage yield of ammonia at 400 °C? [1]

(ii) Explain why the yield decreases as the temperature rises. [2]

AO2 **(b)** In practice, ammonia is produced at 450 °C, even though lower temperatures give a higher yield.

Explain why this temperature is chosen. [2]

AO3 **(c)** The aim of *Green Chemistry* is to use sustainable processes that:

• reduce waste of energy and resources
• involve less hazardous substances.

Explain how the Haber Process to make ammonia could be described as *Green Chemistry*. Give two reasons. [2]

| AO1 | recall the science | AO2 | apply your knowledge | AO3 | evaluate and analyse the evidence |

✱ WORKED EXAMPLE – Higher tier

Most fuel cells being developed today use hydrogen. However, hydrogen is difficult to store and transport. One alternative fuel used in a fuel cell is methanol, CH_3OH.

(a) Draw the full displayed formula for methanol. [1]

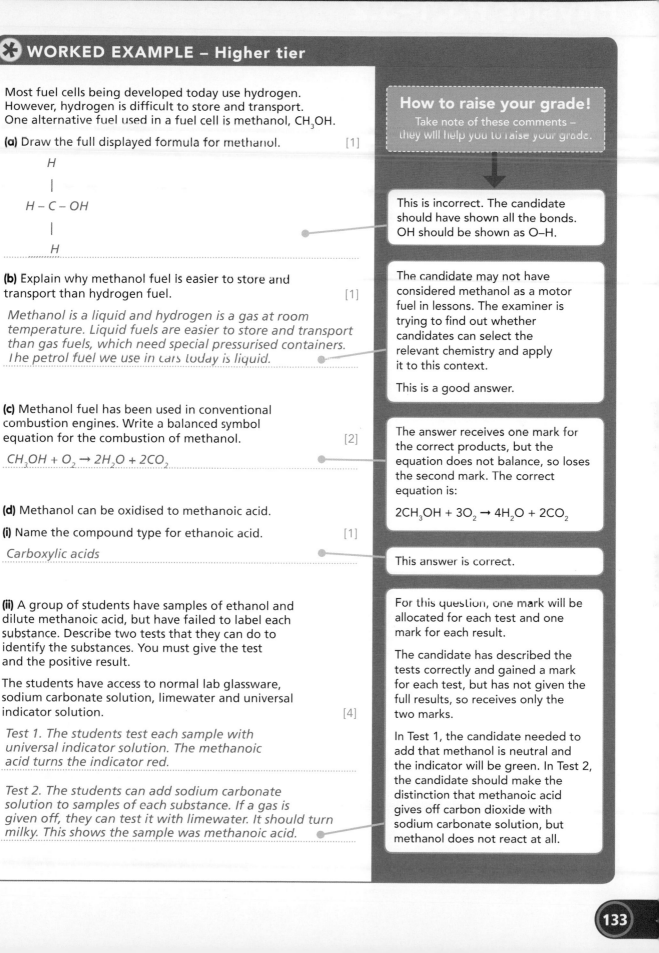

$$\begin{array}{c} H \\ | \\ H - C - OH \\ | \\ H \end{array}$$

This is incorrect. The candidate should have shown all the bonds. OH should be shown as O–H.

(b) Explain why methanol fuel is easier to store and transport than hydrogen fuel. [1]

Methanol is a liquid and hydrogen is a gas at room temperature. Liquid fuels are easier to store and transport than gas fuels, which need special pressurised containers. The petrol fuel we use in cars today is liquid.

The candidate may not have considered methanol as a motor fuel in lessons. The examiner is trying to find out whether candidates can select the relevant chemistry and apply it to this context.

This is a good answer.

(c) Methanol fuel has been used in conventional combustion engines. Write a balanced symbol equation for the combustion of methanol. [2]

$CH_3OH + O_2 \rightarrow 2H_2O + 2CO_2$

The answer receives one mark for the correct products, but the equation does not balance, so loses the second mark. The correct equation is:

$2CH_3OH + 3O_2 \rightarrow 4H_2O + 2CO_2$

(d) Methanol can be oxidised to methanoic acid.

(i) Name the compound type for ethanoic acid. [1]

Carboxylic acids

This answer is correct.

(ii) A group of students have samples of ethanol and dilute methanoic acid, but have failed to label each substance. Describe two tests that they can do to identify the substances. You must give the test and the positive result.

The students have access to normal lab glassware, sodium carbonate solution, limewater and universal indicator solution. [4]

Test 1. The students test each sample with universal indicator solution. The methanoic acid turns the indicator red.

Test 2. The students can add sodium carbonate solution to samples of each substance. If a gas is given off, they can test it with limewater. It should turn milky. This shows the sample was methanoic acid.

For this question, one mark will be allocated for each test and one mark for each result.

The candidate has described the tests correctly and gained a mark for each test, but has not given the full results, so receives only the two marks.

In Test 1, the candidate needed to add that methanol is neutral and the indicator will be green. In Test 2, the candidate should make the distinction that methanoic acid gives off carbon dioxide with sodium carbonate solution, but methanol does not react at all.

> **How to raise your grade!**
> Take note of these comments – they will help you to raise your grade.

Physics P3.1–3.2

What you should know

Sound and ultrasound

Vibrating objects make sounds.

Sound can be reflected, diffracted and refracted.

Very high-pitched sounds are called ultrasound because they are too high for humans to hear.

 Name some animals that communicate using ultrasound.

Light and refraction

Light travels in straight lines from a light source to our eyes.

Light is refracted at the boundary between different materials.

White light is formed from different colours of light. These colours can be separated into a spectrum when the light is refracted.

Our eyes have lenses, which focus light on the retina at the back of the eye.

 Name two pieces of equipment that use lenses.

Pressure, forces and moments

To calculate pressure, you divide force by the area over which the force is applied.

The pressure is larger if the force is applied over a smaller area, making it easier to cut or pierce materials.

Unbalanced forces make objects change shape, speed or direction. They can make objects turn around a pivot.

 Explain why needles have sharp points.

You will find out

Medical applications of physics: X-rays and ultrasound

> X-rays are high-energy electromagnetic waves. They are used to diagnose medical conditions.

> Precautions are needed because X-rays are ionising.

> Ultrasound waves are very high frequency sound waves used in medicine.

> Ultrasound waves partially reflect at boundaries between different materials.

> Oscilloscope traces of ultrasound waves are used to calculate the distance between different materials.

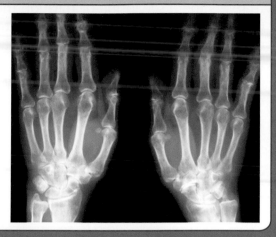

Medical applications of physics: Lenses, the eye and other applications using light

> Lenses use refraction to bring light to a focus and form images of objects.

> Lenses in our eyes allow us to see things clearly.

> Adjusting the shape of the lens or using additional lenses corrects vision defects.

> Total internal reflection allows visible light to travel along optical fibres.

Using physics to make things work

> The time taken for a pendulum to complete a swing depends on its length.

> Objects have a centre of mass, which is where we imagine all its mass is concentrated.

> The moment of a force measures its turning effect. It depends on the size of the force and its distance from a pivot.

> Liquids are incompressible, so forces are transmitted through hydraulic systems. These are used to change the size of forces.

> Objects moving in a circle feel a centripetal force towards the centre, which makes them accelerate by changing their direction. The size of the force can be changed.

Using X-rays

You will find out:

> about the properties of X-rays

> some ways in which X-rays are used

> the precautions taken when using X-rays

Seeing the invisible

Some fictional characters have 'X-ray vision' that lets them see inside or through things. If you have had an X-ray, perhaps on teeth or a suspected broken bone, you will know that X-rays enable medical staff to 'see' inside people. 'X-ray vision', though, would need special equipment because human eyes cannot detect X-rays.

FIGURE 1: This child has swallowed a coin. It would be hard to remove the coin safely, without X-rays to show where it is.

X-rays

Discovery

X-rays were discovered almost by accident. German physicist, Röentgen was experimenting with passing electric currents through a tube of gas (similar to the tube inside early televisions). He found shadows on nearby photographic plates, even when they and the tubes were wrapped in thick black paper.

Now it is known that X-rays are part of the **electromagnetic (EM) spectrum** of waves that includes visible light.

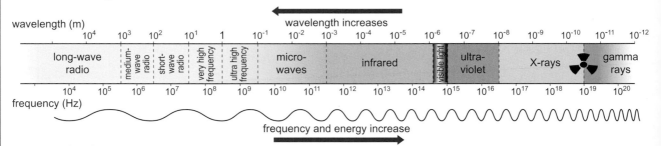

FIGURE 2: The electromagnetic spectrum.

Properties

X-rays:

> travel at the **speed of light** – like all electromagnetic radiation

> have high energy, high **frequency**

> have very short **wavelength** – the same order of magnitude as the diameter of an atom

> cause **ionisation** – by causing electrons to split away from atoms.

They are absorbed by dense materials, such as bone or metals. X-rays are transmitted through less dense materials, such as paper, clothing or healthy tissue.

Because they are ionising, X-rays can damage living cells and can cause cancer.

QUESTIONS

1 X-rays pass through healthy tissue but are absorbed by bone or metal. How does Figure 1 show this?

2 X-rays affect photographic film in the same way as light. Explain how the data in Figure 1 shows this.

Using X-rays

X-rays have many uses in medicine and industry, both for diagnosing and treating problems. X-rays come in a wide range of different energies (all high) and these are useful in different situations.

> The highest energy and most penetrating X-rays are used in industry and research to examine materials.

> Lower-energy X-rays are used in medicine. However, very low-energy rays are unwanted – they are totally absorbed by body tissue – and are filtered out using aluminium sheets.

Medicine

X-rays used to examine bones and teeth have energy high enough to pass through soft tissue and low enough not to pass through bone or teeth.

> Doctors measure bone density using dual-energy X-rays (two different energies). They measure how much of each energy passes through the bone.

> Doctors examine soft tissues using X-rays. Barium and iodine both absorb X-rays. When compounds of barium or iodine are ingested or injected into veins, they make the relevant parts of the body show up on X-ray images.

Computerised tomography (**CT** or **CAT**) **scans**, give a 3-D image of the body. A rotating scanner takes X-ray images from many different directions. A moving bed carries the patient slowly through the scanning machine. A computer combines all the different images, to build up the 3-D image.

Highly focused, energetic beams of X-rays can be used to kill cancer cells.

Displaying X-ray images

X-rays used to be detected using photographic film. Nowadays, digital or computed radiography is used.

The X-rays are detected using charge-coupled devices (CCDs). These turn the X-rays into an electronic image, usually displayed on a computer screen. The CCD contains semiconductor material that is ionised (electrons split away from atoms) by the X-rays. These electrons form a current that can be measured.

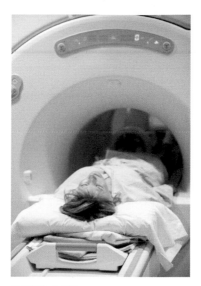

FIGURE 3: CT scanner.

QUESTIONS

3 Why is it a good idea to filter out X-rays that will be totally absorbed?

4 Describe how X-rays can be used to scan soft tissues.

X-rays and safety

X-ray scanners used to be very common in children's shoe shops, to check that shoes fitted properly. They were very popular with the children.

However, all X-rays are potentially harmful. They all increase the risk of cancer, because they are ionising. Although the risk from each individual X-ray is tiny, all unnecessary X-rays should be avoided.

People working with X-ray machines or CT scanners are protected by lead screens to block the X-rays – otherwise they would receive a high overall dose of damaging X-rays. People who work near X-rays wear film badges to monitor their exposure to X-rays.

QUESTIONS

5 Describe how X-rays could be used to check that a shoe fits correctly.

6 Explain why all X-rays are potentially harmful.

Ultrasound

You will find out:
> about ultrasound
> some ways in which ultrasound is used

Bats

Bats are some of the UK's smallest mammals, weighing about the same as a 20p coin. They are amazing creatures. They catch flying insects in total darkness and can even tell the difference between a twig and a twig with a moth on it. How do they do it? Children sometimes claim that they can hear bats squeaking and that is an important clue.

Did you know?

Ultrasound has been used to search Loch Ness for the Loch Ness monster – without definite results one way or the other.

FIGURE 1: Have you ever heard bats squeaking?

Sound

Sound you cannot hear

Humans can hear sounds in the frequency range of about 20 Hz to 20 000 Hz.

As people grow older, they generally become less able to hear high-pitched sounds.

> Young children may be able to hear the squeaks made by bats.

> Older people may be unable to hear time signals or high notes in music.

Sound above the frequency that humans can normally hear is called **ultrasound**.

Ultrasound waves can be produced by electronic systems.

Reflection of sound and ultrasound

Both sound and ultrasound waves can be reflected. The reflection is called an **echo**.

Whenever a sound or ultrasound wave reaches a boundary between two different materials (**media**), some of the wave is reflected.

Echoes can be used to find out the distance to a boundary. Figure 2 shows ultrasound being used to examine the eye. Structures in the eye form boundaries that reflect the ultrasound signal. A computer analyses the reflected signal and builds up a 3-D image of the structures in the eye. From this, ophthalmologists can detect disease or abnormalities.

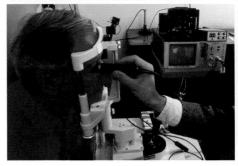

FIGURE 2: Ultrasound being reflected from the eye. Why must the patient not move?

QUESTIONS

1 What is ultrasound?

2 Why are children more likely than adults to hear bats?

3 Explain why a reflected signal is always fainter (quieter, smaller amplitude) than the original pulse.

Using ultrasound

Calculating the distance

Figure 3, on the next page, shows an **oscilloscope** trace for a pulse of ultrasound used to measure the diameter of an eye. The first peak shows ultrasound reflected from the front of the eye. The second peak shows ultrasound reflected from the back of the eyeball.

Q ... human hearing range AND sound longitudinal waves

The horizontal axis of the oscilloscope trace shows time. The distance between the two peaks gives the time t, for the pulse to travel through the eyeball and back.

Sound travels at different speeds in different media. Using v as the speed of ultrasound through the eye, distance, s can be calculated from the equation

$s = v \times t$

Note that distance s is the distance to the back of the eyeball and back. The diameter of the eyeball will therefore be half the distance s.

Medical uses of ultrasound

Ultrasound has several medical uses. The advantage of using ultrasound is that it is not ionising. It does not pose the same risk as X-rays.

> Fetus scans check that a fetus is developing correctly. Different tissues have slightly different densities, so ultrasound is reflected from the boundaries between different tissues. A very detailed image of the developing fetus can be formed.

> Kidney stones and some cancers can be detected.

> Kidney stones can be removed without surgery. A beam of high-energy ultrasound breaks up the stones. The pieces can then be passed harmlessly in the urine.

> **Doppler changes** in an ultrasound signal can monitor blood flow through the heart and major blood vessels.

If the blood is flowing towards the detector, the frequency of the echo will be higher than the original signal. If the blood is flowing away from the detector, the frequency of the echo will be lower than the original signal.

> In 2009, a pioneering treatment used a focused beam of ultrasound to vibrate, heat up and kill cancer cells. It worked as well as conventional treatments, but with far fewer side effects.

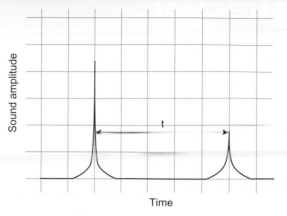

FIGURE 3: Oscilloscope trace for an ultrasound pulse and its echo.

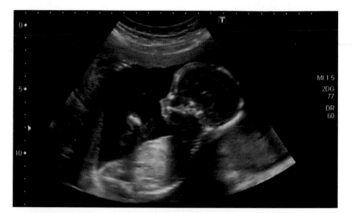

FIGURE 4: Pre-natal ultrasound scan.

QUESTIONS

4 For measuring distances, why is it important to know what material the ultrasound is travelling through?

5 Suggest how the oscilloscope trace shows how much a material absorbs ultrasound.

6 Describe one medical use of ultrasound and how it works.

Skin contact

In an ultrasound examination, the operator will always smear a gel on the skin of the patient before using the ultrasound sensor. This makes the sensor slide smoothly across the skin, but that is not the gel's purpose.

The gel ensures close and continuous contact between sensor and skin. Otherwise, the ultrasound might reflect from the skin, making it harder to obtain a clear image of the internal structures.

QUESTIONS

7 Suggest some properties of the gel used in ultrasound examinations.

Refraction

You will find out:

> what refraction is and why it happens

> how convex and concave lenses form images

Obscure view

The view through some windows is deliberately obscure – the image that you see is distorted. The glass has a different thickness in different places. Rays of light passing through the window are bent to a different extent. Plain windows in old houses give slightly distorted images because it used to be much harder to make large, flat panes of glass.

FIGURE 1: Suggest why some front doors have this type of window.

Refracting light

Refraction is the change in direction of light as it passes from one medium (material) to another. Refraction makes a stick in water appear bent. It can make a coin in the bottom of a mug appear to disappear.

Figure 2 shows that:

> refraction happens at the boundaries between different materials – it does not happen inside the glass block itself

> light refracts (changes direction) towards the **normal** as it goes into a more dense medium (from air to glass)

> light bends away from the normal as it goes into a less dense medium (glass to air).

Why refraction occurs

It is harder for light to travel through glass than through air. Light slows down a little as it goes into glass and speeds up again as it leaves.

> If a light ray hits a glass block at an angle, one side of the ray will slow down first. The ray will bend towards the block, towards the normal.

> If a light ray is travelling along the normal, the change in speed does not make it change direction.

How lenses work

Light is refracted (changes direction) at each surface of a **lens**. Figure 3 shows how the light rays refract.

Each light ray hits the lens at a slightly different angle. Each ray is refracted differently from its neighbour.

> A **convex** lens is a **converging** lens – the rays of light come together.

> A **concave** lens is a **diverging** lens – the rays of light spread out.

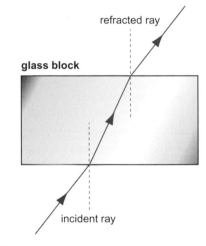

glass block

FIGURE 2: Light passing from air to a glass block and back out to air.

QUESTIONS

1 Which way is light refracted as it passes from glass to air?

2 When light goes from air to glass, what happens to its speed?

3 Water refracts light in exactly the same way as glass. What does this tell you about the speed of light in water?

Convex   **Concave**

FIGURE 3: Light rays as they pass through lenses.

The power of lenses

Conventions

Figure 4 shows how lenses are represented on drawings. Light is shown as only changing direction at the centre line of the lens.

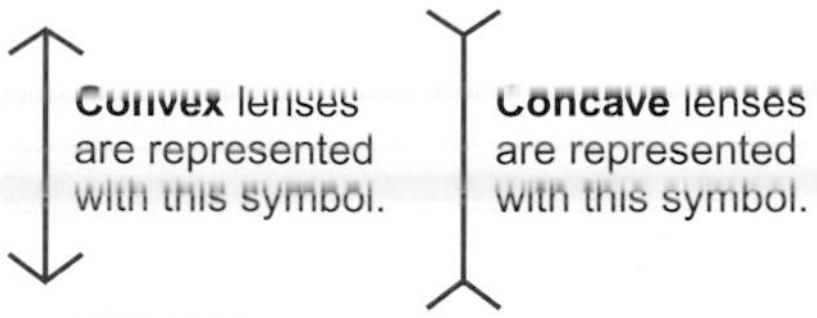

Convex lenses are represented with this symbol.

Concave lenses are represented with this symbol.

FIGURE 4: Lens symbols.

Figure 5 shows what happens when parallel rays of light pass through convex or concave lenses.

The **principal focus**, F, for a lens is the point where parallel rays of light are brought to a focus.

The **focal length**, f, of the lens is the distance from the centre of the lens to the principal focus, F.

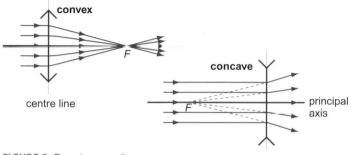

FIGURE 5: Drawing ray diagrams.

Refractive index

The **refractive index** of a medium indicates how much the light changes direction as it enters or leaves that material. The greater the refractive index, the more the light changes direction.

$$\text{refractive index} = \frac{\sin i}{\sin r}$$

where
i is the angle of incidence
r is the angle of refraction

Note that these angles are measured between the light ray and the normal.

Calculating power

A powerful lens is one that has a big effect on the light going through it. The more powerful the lens, the more the light changes direction – the focal length will be shorter.

The **power of a lens** is found from the equation:

$$P = \frac{1}{f}$$

where
P is the power in **dioptres** (D)
f is the focal length in metres (m)

The power is said to be positive for a convex (converging) lens and negative for a concave (diverging) lens.

Focal length depends on three factors.

1. How curved the lens is – the greater the curvature, the more light is refracted and the shorter the focal length.

2. The refractive index of the material that the lens is made from – a higher refractive index, the more the light is refracted and the shorter the focal length.

3. The medium that the lens is in – a lens in air refracts more than the same lens in water.

QUESTIONS

4 Describe how you could measure the focal length of a concave lens.

5 Suggest a way to find the refractive index for glass using a rectangular block of glass.

6 Calculate the power in dioptres of a lens with a focal length of 10 cm.

Making thinner lenses (Higher tier)

For a given focal length, the greater the refractive index of the material, the flatter the lens can be.

By using materials with a higher refractive index, lenses can be made thinner and lighter. This is useful for lenses in optical equipment where weight is important.

QUESTIONS

7 Glass has a refractive index of 1.52 and plastic has a refractive index of 1.49. Describe the differences that you would expect to see between a glass lens and a plastic lens of identical size and shape.

Q ... refractive index GCSE

Preparing for assessment: Planning an investigation

To achieve a good grade in science, you not only have to know and understand scientific ideas, but you need to be able to apply them to other situations and investigations. These tasks will support you in developing these skills.

✴ Investigating the refractive index of jelly

Laura was asked to investigate how the concentration of jelly affects its refractive index.

Method

Samples of solid jelly were made by dissolving different amounts of jelly cubes from the same packet in 100 cm³ of water and leaving the jelly to set.

When the jelly was set, Laura cut it into pieces of exactly the same size.

She shone a narrow ray of light into each block of jelly.

Laura measured the angle of incidence and the angle of refraction.

Using this information, Laura could calculate the refractive index of each sample.

✴ Planning

1. Write a hypothesis linking the refractive index of jelly and the concentration of the jelly.

Always identify the independent variable and dependent variable before writing a hypothesis.

2. Write down two variables that must be controlled and how to control them.

Include the variables that have the most effect on the results if they are not controlled properly.

3. Explain what might limit the range of measurements that Laura would make.

Your method should aim for as wide a range of readings as possible, given the practical considerations.

4. Write down a planned method for this experiment.

Your method should include all the steps to obtain a full set of results, including the equipment that you plan to use.

✳ Processing data

5. Prepare a table for your results.

6. Explain how you will use your results to calculate the refractive index.

> Your table should include space for calculations involving the results as well as the actual results.

> Remember that the refractive index for any one block should be the same, regardless of the angle of incidence.

✳ Assessing and managing risks

7. List any risks that needed to be controlled at different stages in your experiment, and how these can be reduced.

> The risks are different for different stages so think carefully about the whole experiment.

✳ Reviewing the investigation

8. Some other students carried out a similar experiment. They found that different angles of incidence gave them different values for refractive index.

Suggest how the students could make sure which result is correct.

> For this experiment, it is worth thinking about how to repeat results because the original sample will be discarded before the students can analyse their results.

✳ Connections

How Science Works

- Plan practical ways to develop and test scientific ideas
- Assess and manage risks when carrying out practical work
- Collect primary and secondary data
- Select and process primary and secondary data

Science ideas

P3.1.3 Lenses

Lenses

Larger or smaller

You have almost certainly used lenses for something. If you have taken a photograph, the camera had a lens in it – even if it was on your phone. Perhaps you have also used a magnifying glass or microscope to make small things look larger, or binoculars or a telescope to see things far away.

> **You will find out:**
>
> > how to draw ray diagrams for light passing through convex and concave lenses
> > how to calculate magnification
> > how to describe images

FIGURE 1: What type of lenses are there in a microscope?

Drawing ray diagrams

Objects look different when viewed through different lenses: different lenses form different images.

A **ray diagram** is a diagram drawn to scale to find the position and size of an image formed by a lens. Ray diagrams are constructed in the same way for both concave and convex lenses.

1. Select a suitable scale.

2. Draw the **principal axis**, the centre line of the lens, and the lens. Mark C, the centre point of the lens.

3. Mark F, the focus. This is at the focal length from C.

4. Draw the object (the correct height) at the correct distance from the lens. Measure the distance from the centre line of the lens.

5. Draw a ray from the top of the object straight through point C on the lens.

6. Draw a ray from the top of the object to the lens, parallel to the principal axis, then through to the principal focus on the other side of the lens.

7. These two rays cross at the top of the image.

If you have to extend one or more rays backwards to find where they cross, draw these as dotted lines (they are '**virtual**' or 'imaginary' rays).

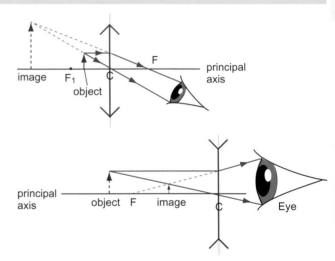

FIGURE 2: Ray diagrams for a convex lens and a concave lens.

QUESTIONS

1 An object is 2 cm tall and placed 2 cm from a convex lens that has a focal length of 4 cm. Draw the ray diagram. Complete the rays backwards as dotted lines to draw the image.

Magnification

The **magnification** of a lens indicates how much it magnifies an image – how much larger or smaller than the object it appears.

The magnification of a lens can be calculated from the equation:

$$\text{magnification} = \frac{\text{image height}}{\text{object height}}$$

Did you know?

You can use a tiny drop of water on a sheet of cling film as a powerful magnifying lens.

> A magnification of 1 means that the image is the same size as the object.

> A magnification of less than 1 means that the image is smaller than the object.

Figure 3 shows that a lens with a short focal length makes a better magnifying glass because the magnification is greater and, therefore, the image is larger.

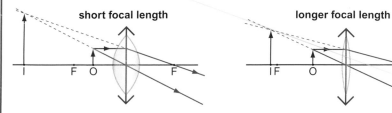

FIGURE 3: Both lenses are convex, but thickness matters.

> **QUESTIONS**
>
> **2** A magnifying glass has a magnification of 6. When looking at a pin 0.05 mm wide, what would the width appear to be?
>
> **3** Using a lens, an insect that is 4 mm wide looks 6 mm wide. What is the magnification of the lens?

Images

The type of image formed in a convex (converging) lens depends on the distance between the object and the lens. The ray diagrams in Figure 4 show the image formed when an object is placed closer than the focal length, more than the focal length and more than twice the focal length (2F).

The ray diagrams in Figure 5 show that a concave (diverging) lens always forms the same type of image, wherever the object is placed.

Describing images

When images are formed by one or more virtual, or imaginary, rays – drawn as dotted lines – they are **virtual images**. If you put a screen in the place where the image is, you would not see anything on the screen.

To describe an image you have to say if it is:

> virtual or real

> upright or **inverted** (upside down compared with the object)

> magnified (larger), same size, or smaller than the object – or give the size of the magnification.

The image for a concave lens is always virtual, upright, and smaller than the object.

The image for a convex lens can be virtual or real, upright or inverted, magnified, the same size or smaller than the object.

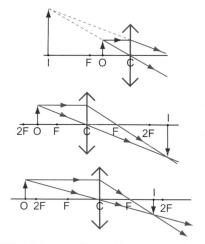

FIGURE 4: Images formed by a convex lens.

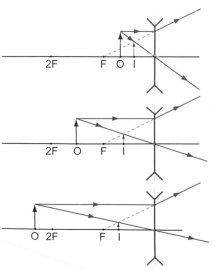

FIGURE 5: Images formed by a concave lens.

> **QUESTIONS**
>
> **4** If an image is 'virtual', what does this mean?
>
> **5** Where must you place the object so that a convex lens gives an image with a magnification of more than 1?

Seeing clearly

Spectacular sight

Over half the people in the UK wear spectacles or contact lenses – and many more need to. There is a wide range of different lenses available to suit different people's needs. Lenses can correct long sight, short sight, astigmatism and other problems. A lens can be complex with several curvatures in the one lens. There are many options.

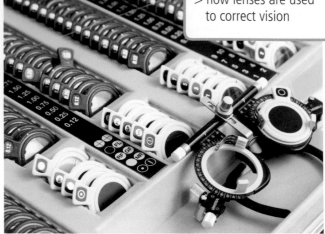

FIGURE 1: A lens set that an optometrist uses, in one part of an eye examination, contains 228 lenses that may be combined in over 20 000 ways.

The eye

The eye is truly amazing. The **range of vision** – where a person with normal vision sees things clearly – goes from the **near point**, approximately 25 centimetres from the eye, to the **far point**, at infinity.

The eye's structure

> The **cornea** is a clear surface that allows light through but protects the eye from dust and infection. About two-thirds of focusing by the eye happens here.

> The **pupil** is where light enters the eye. The iris changes size to make the pupil larger (in dim light) or smaller (in bright light). Enough light enters to form a clear image, but not enough to damage the eye.

> The **lens** changes shape to 'fine-tune' the light to focus on the retina.

> The **retina** has light-sensitive cells that send electrical signals along the optic nerve to the brain. The brain 'interprets' the pattern of light on the retina, so that you see an image.

> The **suspensory ligaments** hold the lens in place. They attach to the **ciliary muscles** which pull on the lens to change its shape. The lens is thicker to focus on near objects and flatter to focus on distant objects.

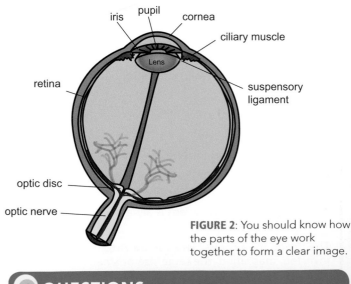

FIGURE 2: You should know how the parts of the eye work together to form a clear image.

QUESTIONS

1 Give values for the distance, from the eye, of (a) the near point (b) the far point.

2 Describe how you would expect to see someone's pupil change, if they went from somewhere brightly lit to somewhere darker.

3 What is the function of the retina?

Q ... the eye GCSE

Comparing an eye with a camera

The basic structure of any camera is very similar to the eye, but has some important differences.

In all old cameras, the light-sensitive surface was photographic film. Each piece of film could be used only once. **Digital cameras** have a charge coupled device (CCD) as a detecting surface, which is more similar to a retina. Light falling on the semiconductor surface causes an electric charge. A computer records the charge at each point (each pixel) of the surface and converts this into a 'picture' that can be recorded, sent or displayed electronically.

In a camera, the image is focused by moving the lens backwards and forwards to change the distance between the lens and the detecting surface. Unlike the eye's lens, a camera lens cannot change shape.

When you take a photograph, the camera has to 'freeze' the light reaching the detecting surface. The camera shutter allows light onto the detecting surface for the right length of time to form a good image – not too bright or too dark. In a digital camera this can be controlled automatically.

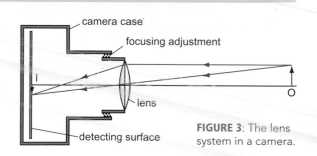

FIGURE 3: The lens system in a camera.

● QUESTIONS

4 Explain what happens when you focus a camera.

5 Compare an eye with a camera by making two lists, headed Similarities and Differences.

Correcting vision

Many older people need to wear spectacles when reading. The relaxed position for the eye is a flatter lens, for distance vision. With age, the lens does not change shape as easily, so people become unable to see close things clearly.

People of any age can be **short sighted** or **long sighted** because of the shape of their eye.

Short sight

> People cannot see distant objects clearly.

> The eyeball is too long and the image forms in front of their retina.

> This is corrected by a concave lens, which makes rays of light from a distant object diverge.

> The image forms further back, on the retina.

Long sight

> People cannot see near objects clearly.

> The eyeball is too short and the image forms behind their retina.

> This is corrected by a convex lens, which makes rays of light from a near object converge.

> The image forms further forward, on the retina.

Lighter light (Higher tier)

Spectacles can be made lighter and so more comfortable to wear, by using lenses with a high refractive index. This means that the lens can be thinner while having the same effect on the light.

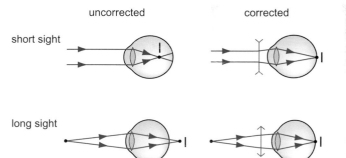

FIGURE 4: Correcting short sight and long sight.

● QUESTIONS

6 Describe what causes short sight, and how it is corrected.

7 Use diagrams to explain how spectacle lenses with a higher refractive index can be thinner, yet still give the same vision correction.

More uses of light

Decorative lighting

Optical fibre lights contain bendy strands of glass. The strands look just like white plastic from the outside but transmit light along the inside to shine out from the end. Decorations look nice, but optical fibres have many, much more important uses.

FIGURE 1: Can you think of serious uses for fibre optics?

Applications using light

Optical fibres are made from bendy glass. If glass is stretched quickly while it is molten, it forms thin, flexible strands. The strands will snap if you try to bend them too far.

Light shone in one end of the optical fibre reflects back and forth down the fibre until it comes out the other end.

Endoscopes

Medical specialists use optical fibres to examine inside patients without the need for surgery.

An **endoscope** is a thin, flexible tube. It contains an optical fibre and cables to a camera at its tip. The endoscope is inserted in the patient. The camera and light transmitted along the optical fibre enable a doctor to make a visual examination.

Sometimes, tiny surgical instruments are used, for example to take a biopsy.

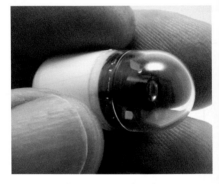

FIGURE 2: Cameras used with endoscopes are small enough to swallow.

Lasers

Lasers produce a very narrow, highly focused, high-energy beam of single wavelength electromagnetic radiation, often light. Even a relatively low-energy laser, such as a laser pointer, can permanently damage your eyes if looked at directly.

Lasers deliver a beam of very high energy radiation to a very small area. Therefore, lasers have many applications where very precise cutting or burning is needed. They are used in industry, research, by the military and in medicine.

Medical uses of lasers include:

> cosmetic surgery to remove tattoos by breaking up the ink particles

> general surgery, to make incisions by vaporising tiny regions of cells,

> general surgery, to cauterise blood vessels, reducing bleeding and swelling after surgery

> eye surgery, to repair torn or detached retinas by welding the tissue back together

> correction of vision, by using precise laser cutting to reshape the cornea

> cancer treatment, to destroy tumour cells.

FIGURE 3: Many, but not all, lasers produce visible light.

QUESTIONS

1 Describe how an optical fibre works.

2 Why does an endoscope need a light and a camera?

3 What features of lasers make them useful in so many ways?

4 Research one medical use of lasers and one non-medical use.

Did you know?

Lasers are given a *Class* label according to their power and wavelength. The higher the class number, the more dangerous the laser.

Total internal reflection

Light is refracted at any boundary between two media. If the angle of incidence is large, the light will not pass the boundary. It is reflected back instead. Figure 4 shows what happens as the angle of incidence increases for a ray of light leaving a glass block.

> In diagram A, the light ray refracts (changes direction) away from the normal as it passes from the glass block into the air. As the angle of incidence increases, the angle of refraction also increases.

> In diagram B, the angle of refraction has increased to 90°. The angle of incidence when this happens is the **critical angle**.

> In diagram C, the angle of incidence is greater than the critical angle. All the light is reflected back into the glass block. This is **total internal reflection**.

QUESTIONS

5 What does 'critical angle' mean?

6 When a light passes into a glass block there is always some reflection. Redraw diagrams A and B in Figure 4 and show the direction of reflected light in each case.

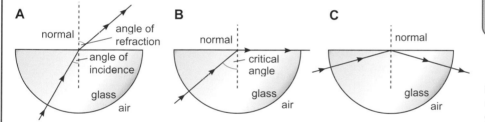

FIGURE 4: Total internal reflection in a semicircular glass block.

Optical fibre

Optical fibres have a high-density glass core, surrounded by a lower-density glass cladding. When a light ray hits the boundary between the inner core and the cladding, it is totally internally reflected back into the core. The light ray continually reflects back and forth, within the fibre, until it emerges from the other end.

Critical angle (Higher tier)

The refractive index (*n*) of a medium (material) is found from the equation:

$$n = \frac{\sin i}{\sin r}$$

This is measured for a ray of light going from air to a denser medium.

In the measurement of critical angle, light goes the opposite way, from glass to air. The equation still works, but the values of *i* and *r* must be corrected for the ray going the opposite way.

$$n = \frac{\sin 90°}{\sin c}$$

where *c* is the critical angle.

This gives the equation $n = \dfrac{1}{\sin c}$

QUESTIONS

Critical angles:
for quartz: 40.5°
for diamond: 24.5°

7 A ray of light strikes the inside of a diamond block at an angle of incidence of 45°. Will the ray emerge from the block?

8 Calculate the refractive index of (a) quartz (b) diamond.

Preparing for assessment: Applying your knowledge

To achieve a good grade in science, you not only have to know and understand scientific ideas, but you need to be able to apply them to other situations and investigations. These tasks will support you in developing these skills.

✹ Bionic eyes

A damaged retina can lead to permanent, untreatable blindness for some unlucky people. Causes of damaged retinas include disease, genetic defects and accidents. Once the retina stops working properly, the sight in that eye becomes much worse. The retina contains light-sensitive cells that process the light that falls on them. Electrical impulses are sent from these cells to the brain, which interprets the impulses as images. If too many cells in the retina are damaged, it is not possible to repair them and restore sight.

A scientific breakthrough has given hope to some people with particular types of damage. An external video camera is attached to the patient's head. This films the images that the person would see if their eyes were working normally. A small computer changes the images from the video into a series of electrical impulses. These impulses are transmitted wirelessly to an electrode array implanted in the eye. The electrode array processes the signals and sends them to the brain to be interpreted.

At the moment, the implants contain about 60 electrodes. This gives a very limited vision as each electrode generates a single dot in the image in the brain. If we view our surroundings as just 60 changing dots, it would be possible to make out only outlines and basic shapes. To make out a human face accurately, you would need an array with 1000 electrodes. These are the types of improvements that the scientists working on the artificial retina are trying to make.

If other parts of the eye are damaged , it causes serious visual problems. However, the retina contains the specialist cells that are so hard to replace artificially.

A magnified cross section of the retina.

The area of the human retina is about one square centimetre.

The retina has 125 million of these photoreceptors.

✸ Task 1

The scientists used an external video camera to help people see.

(a) The structure of any camera has several features in common with the eye.

Write down three ways in which a camera and an eye are similar.

(b) The camera focuses on images at different distances by moving the lens.

Explain how the lens in the eye focuses on images at different distances.

✸ Task 2

A retina working normally has about 125 million cells. The artificial retina has about 60 electrodes.

(a) Explain why a healthy retina is so important for vision.

(b) How does the number of cells or electrodes affect the detail that a person can see?

✸ Task 3

A very common eye problem is a cataract. The lens of the eye becomes cloudy.

This problem is treated by removing the lens of the eye and replacing it with a clear artificial material, such as plastic or silicone.

(a) Explain why a cataract causes problems with eyesight.

(b) Treating a cataract is simpler than treating problems with the retina. Explain why.

(c) People who have had a cataract operation need a new prescription for their spectacles. Explain why.

✸ Maximise your grade

	Answer includes showing that you...
E	know the names of different parts of the eye.
	recall that the eye and camera have similarities and differences.
	recall that lenses can correct vision.
	can describe the function of different parts of the eye.
	recall which types of lens correct different vision problems.
C	can compare the structure of an eye with a camera.
	can compare the function of different parts of an eye and a camera.
	can explain how lenses correct vision.
A	can evaluate the use of lenses to correct vision defects.

Centre of mass

You will find out:

> about the centre of mass of an object
> how to find the centre of mass of a thin sheet of material
> the equation for a pendulum that connects periodic time and frequency

Balancing act

Have you ever tried standing with your back to a wall, with your heels touching the wall, and then tried to pick up something from the floor just in front of your feet? Can you stand upright beside a wall, with arms by your sides and both feet together, and the side of one foot touching the wall? Can you explain what happens?

Did you know?

The long pole that tightrope walkers carry is not for show. The pole lowers their centre of mass, making it easier for the acrobat to balance.

FIGURE 1: What happens if this girl moves her feet towards the wall?

Centre of mass and stability

Centre of mass

Imagine all the mass of an object collapsing inwards to the middle of the object. The **centre of mass** is the point where all the mass would end up. It is where the mass of the object may be thought to be concentrated.

If you suspend any object, it will come to rest with its centre of mass directly below the point where it is suspended.

object suspended from any point of triangle

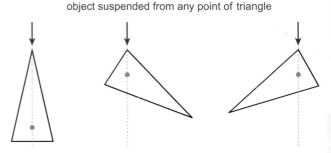

FIGURE 2: No matter where you suspend a triangular piece of card, it always comes to rest with its centre of mass (●) directly below where it is suspended.

TABLE 1: Examples of centre of mass.

Object	Centre of mass
sphere	centre of the sphere
ruler	halfway along its length and halfway across its width
cone	much nearer its wide end than the point
any symmetrical object	on its axis of symmetry

Stability

Stability indicates how likely it is that an object will topple over. Stable objects are those that do not topple over easily. The more unstable an object is, the more easily it topples. The stability of an object depends on its shape and the position of its centre of mass.

FIGURE 3: An object is more stable if it has a wide base. The vase on the left is less likely to topple than the one on the right.

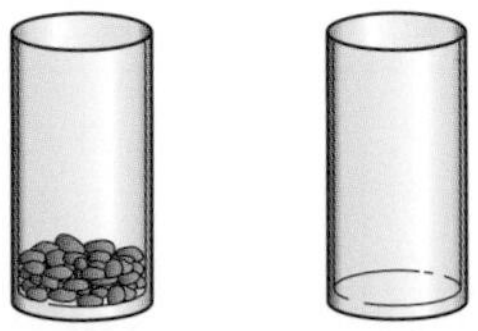

FIGURE 4: An object is more stable if its centre of mass is lower. The vase on the left is less likely to topple.

Q ... finding centre mass uniform bodies

Remember
Whatever the object, its centre of mass will be directly below the point of suspension.

QUESTIONS

1 An object is unstable. Describe in your own words what this means.

2 State the two factors that affect the stability of an object.

3 Suggest how you would design a lorry to make it as stable as possible.

Finding the centre of mass

A freely suspended object always comes to rest with its centre of mass directly below its point of suspension. Figure 5 shows how to find the centre of mass of any thin sheet, no matter how irregular its shape may be.

> Suspend the sheet of material in turn from three different holes near its edge.

> For each hole, use a plumb line to find the vertical line running down from the point of suspension.

> The centre of mass of the sheet is where the three lines cross.

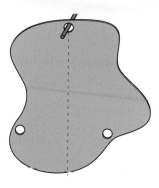

FIGURE 5: Finding the centre of mass.

QUESTIONS

4 (i) Suggest where the centre of mass of a car tyre is found.

(ii) Describe an experiment to find out if you are correct.

Pendulums

If you move a suspended object so that its centre of mass is no longer below the point of suspension, it will begin to swing. The object will swing to and fro until it final reaches its starting position. It is behaving as a **pendulum**.

If a pendulum is set swinging, the time it takes for one complete swing is the **periodic time**. For a simple pendulum

$$T = \frac{1}{f}$$

where
T is periodic time in seconds (s)
f is frequency in hertz (Hz)

A playground swing is an example of a pendulum, as are other simple fairground and playground rides.

Example

A swing has been pulled back and then released. If its frequency is 0.25 Hz, how long will it take the swing to return to where it was released?

$$T = \frac{1}{f}$$

$$T = \frac{1}{0.25} = 4 \text{ seconds}$$

The time period of a pendulum depends on its length – the distance from its point of suspension to its centre of mass. It does not depend on the mass or amplitude.

QUESTIONS

5 A pendulum takes 42 seconds to complete 6 swings. Calculate its frequency.

6 Two identical balls are each suspended on a piece of string. One piece is twice the length of the other. Both are pulled back to the same distance from the vertical. Suggest how the frequencies of the two pendulums might differ.

Moments

Lifting bridges

Lifting bridges are quite common across some canals. The bridge is strong enough to drive a herd of cows over and weighs several tonnes. Yet, one person can lift the bridge, to allow a narrow boat to pass underneath.

FIGURE 1: How can one person lift this bridge? Moments give the answer.

What are moments?

Forces can make things turn. For example, a turning force opens a door or undoes a nut. The turning effect that a force has is its **moment**.

> Pushing open a heavy door, you have to push hard. The moment is larger if a larger force is used.

> Trying to undo a really tight nut is easier with a longer spanner. The moment is larger if the distance between the force and the **pivot** (or axis of rotation) is longer.

The size of the moment is calculated using the equation:

$M = F \times d$

where
M is the moment of the force in newton metres (N m)
F is the force in newtons (N)
d is the perpendicular distance from the line of the action of the force to the pivot in metres (m)

FIGURE 2: These valves control water flow in pipes. Which do you think is easiest to turn?

QUESTIONS

1 What is the turning effect of a force called?

2 What two factors make the turning effect of a force larger?

3 Calculate the size of the moment of a 30 N force acting 50 cm from a pivot.

Balanced or unbalanced

Each person on a seesaw exerts a force that acts at a distance from the pivot.

the size of each moment =
    the force the person exerts (the weight) × their perpendicular distance from the pivot

The moment of a person at one end of a seesaw tries to turn it clockwise. The person at the other end has a moment that tries to turn it anticlockwise.

For seesaw A, in Figure 3, the moments are the same size because:

> the forces are the same size (both masses have the same weight)

> the distance of each from the pivot is the same.

The seesaw is balanced.

For seesaw B, the total clockwise and anticlockwise moments are not balanced.

An object is unbalanced if the moment in one direction is greater than the moment in the opposite direction. There is a **resultant moment**.

A
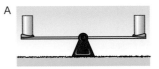

B

C

FIGURE 3: Two balanced seesaws (A and C) and one unbalanced seesaw.

Levers

A **force multiplier** reduces the force needed to move an object.

A seesaw is an example of a simple **lever**. The unbalanced seesaw B in Figure 3 can be balanced if the heavier mass moves nearer to the pivot, as in seesaw C. This is a force multiplier.

Alternatively, the pivot could be moved, as shown in Figure 4.

The force multiplier has the greatest effect when:

> the force you are trying to overcome is as close to the pivot as possible, to decrease its moment

> the force you are using is as far from the pivot as possible, to increase its moment.

FIGURE 4. By moving the pivot, the seesaw becomes a force multiplier. A small child could lift a heavier adult

Balancing moments (Higher tier)

Objects are balanced – they do not turn – when the clockwise and anticlockwise moments are balanced. Knowing this, you can calculate the size of the force needed to stop something turning.

$F_1 \times d_1 = F_2 \times d_2$

Example

Calculate the force needed to stop the ruler in Figure 6 rotating, when it is suspended from its centre point.

clockwise moment = $300 \times 0.4 = 120$ N m

For the ruler to balance:

anticlockwise moment = clockwise moment = 120 N m

$120 = X \times 0.2$

Therefore $X = 120 \div 0.2 = 600$ N

FIGURE 5. Calculating balanced moments.

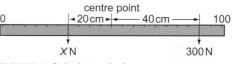

FIGURE 6: Calculating balance.

QUESTIONS

4 A ruler is suspended from its centre point. A 50 g mass is hung 40 cm from the pivot. Calculate how far from the other side of the pivot you would have to put a 20 g mass, to make the ruler balance.

Tendency to topple (Higher tier)

The line of action of the weight of an object runs vertically from its centre of mass. If you try to topple it, the pivot is the point at which the object tries to turn.

> If the line of action passes through the object's base, the object will not topple.

> If the line of action does not pass through the object's base, there will be a resultant moment. The object will topple.

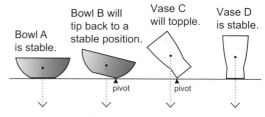

FIGURE 7: Why objects may topple.

QUESTIONS

5 Use moments to explain why racing cars are constructed so that they are low to the ground.

Hydraulics

You will find out:
> how force is transmitted through a liquid
> how hydraulic systems can be used as force multipliers

Deep-sea submersibles

The deeper that deep-sea submersibles are designed to go, the more rounded they become. The pressure increases as you go deeper beneath the ocean and a round shape is stronger – it is less likely to be crushed. The force acts on the submersible in all directions, so the submersible has to be strong in all directions.

FIGURE 1: *Nautile* carries out underwater research at depths of six kilometres.

Pressure and force in liquids

Pressure

It is almost impossible to **compress** a liquid. You can show this by trying to squash a plastic bottle filled completely with water.

Water inside a container would flow out if it could. It exerts a **pressure** on the walls of the container.

In Figure 2, water comes out of the plastic bag, wherever there is a hole. If you squeezed one part of the bag – increasing the pressure at that point, water comes out faster from all the holes. Pressure in a liquid acts in all directions, forcing the water out.

Hydraulic systems

Hydraulic systems use liquids in pipes. Because liquids are incompressible, hydraulic systems can transmit force from one place to another.

In Figure 3, pushing on the first syringe (*force in*) results in a *force out* from the second syringe. The pressure transmits the force through the liquid.

This type of system is used in car braking systems.

Force multipliers

A lever is a force multiplier – it increases the distance between force and pivot. This decreases the force needed to move something.

Hydraulic systems can also be used as force multipliers.

If the area that the *force in* pushes on is smaller than the area for the *force out*, the force will multiply.

To have as large a *force out* as possible, there should be:

> a small area for the *force in*

> a large area for the *force out*.

FIGURE 2: Pressure in a liquid acts in all directions.

force in

force out

FIGURE 3: A simple hydraulic system.

QUESTIONS

1 Describe what 'incompressible' means.

2 How can you show that liquids are incompressible?

3 State two facts about pressure in a liquid.

4 Explain how a hydraulic system can be used as a force multiplier.

Calculating the force

The pressure at any point in a hydraulic system is found from:

$$P = \frac{F}{A}$$

where
P is the pressure in pascals (Pa)
F is the force in newtons (N)
A is the cross-sectional area in metres squared (m^2)

The hydraulic system in Figure 4 is being used as a force multiplier.

The pressure at the left-hand cylinder is:

$$P = \frac{F1}{A1}$$

where A1 is the cross-sectional area of the left-hand cylinder.

The pressure is the same at all points in a hydraulic system. Therefore,

$$P = \frac{F2}{A2}$$

where A2 is the cross-sectional area of the right-hand cylinder.

Combining these two equations gives:

$$\frac{F1}{A1} = \frac{F2}{A2}$$

Therefore $F2 = \dfrac{(F1 \times A2)}{A1}$

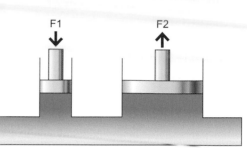

FIGURE 4: Cylinders in a hydraulic system. How are the forces F1 and F2 related?

QUESTIONS

5 A cylinder with a cross-sectional area of 0.0001 m^2 is connected to a brake pedal. A driver pushes down with a force of 100 N on the pedal. The cross-sectional area of the cylinder connected to the brake pads is 0.1 m^2. What will the force be, on the brake pads?

FIGURE 5: Hydraulic disc brakes give mountain hikers huge stopping power.

Conservation of energy

It looks as though a hydraulic system using a small force to produce a very large force breaks the Law of Conservation of Energy. It does not. The energy transferred is the work done by a force.

work done = force × distance moved in direction of force

The smaller force moves through a large distance, while the larger force moves through a small distance. It may help to imagine where the liquid would go if both pistons moved the same distance.

QUESTIONS

6 A braking system multiplies the applied force by 100. If the brake pedal moves through a distance of 5 cm, how far will the brake pads move?

Circular motion

You will find out:
> the force and acceleration acting on an object moving in a circle
> factors affecting the centripetal force needed to keep an object moving in a circle

Hammer throwing

Hammer throwing is a popular sport in Scottish Highland Games. An athlete whirls the heavy 'hammer' round and round, and then releases it. The winner is the athlete whose hammer goes furthest, in the right direction. What forces do you think act on the hammer and on the athlete?

FIGURE 1: Why is the hammer whirled round before it is released?

Circular motion

Circular motion and force

If an object changes **speed** or **direction**, there is a **resultant force** on it. An object moving in a circle, with a constant speed, is changing direction all of the time – there must be a force acting on the object all of the time. Without this force, the object would fly off in a straight line, like the hammer does when the hammer thrower lets go of it.

This force is **centripetal force**. It always acts towards the centre of the circle.

Circular motion and acceleration

An object with a resultant force acting on it **accelerates** in the direction of the force.

This acceleration makes its **velocity** change.

When an object moves in a circle with a constant speed, its velocity is changing. Velocity is speed *in a particular direction*. Therefore, if the object's speed is constant, its direction is constantly changing.

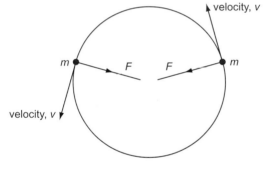

FIGURE 2: Centripetal force F, acting on mass m.

Remember
Centripetal force always acts towards the centre of the circle.

QUESTIONS

1 What tells you that there a resultant force on an object moving in a circle?

2 For an object moving in a circle, is there a change in (a) speed (b) velocity (c) direction?

3 Explain how an object moving in a circle can be accelerating, even when its speed is constant.

Did you know?

The Earth orbits the Sun at a speed of almost 30 kilometres per second. We do not notice this speed because we are not speeding up or slowing down.

Centripetal force

How large is the centripetal force?

Any moving object continues moving in a straight line unless a resultant force acts on it.

The **centripetal force** is the force needed to stop the object flying off in a straight line.

Compare the action in Figures 1 and 3. The athlete needs to exert centripetal force to hold onto the shaft or cable. What affects this 'pull' that the athletes feel?

> The force increases as the mass increases – a massive hammer is harder to whirl than a light mass.

> The force increases as the speed increases – as a hammer goes faster, it becomes harder to hold on to.

> The force increases as the radius of the circle decreases. For any given speed, the smaller the circle the more rapidly the direction is changing. A larger force is needed to make this change happen.

What provides the centripetal force?

The **tension** in the hammer cable, in Figure 3, provides the centripetal force needed to keep the mass moving in a circle. Sometimes, though, you cannot see what is providing the centripetal force.

> For satellites and planets, the force due to gravity provides the centripetal force.

> For vehicles driving around bends, **friction** between the tyres and the road provides the centripetal force. On race tracks the bends are banked so that a component of the weight of the vehicle acts down the slope and adds to the force due to friction. This increases centripetal force and enables cars to drive faster round the bend.

FIGURE 3: What factors affect the centripetal force needed here?

> ### QUESTIONS
>
> **4** State two factors that would reduce the centripetal force needed by a hammer thrower.
>
> **5** What would happen if the friction on a road bend was not great enough to provide the centripetal force needed?
>
> **6** Explain why a fast car is more likely to skid off the road at a sharp bend than a slow car.

Centrifuges

Centrifuges are used to separate liquids that are mixtures of components with different masses. As the centrifuge tube is spun, the heavier particles move to the end of the centrifuge tube.

Circular motion explains why this happens.

As the tube in Figure 4 is spun, at a given speed, the centripetal force is just strong enough to keep a particle of mass m in circular motion. It will not be strong enough to keep a particle of larger mass M in circular motion. Particles of larger mass tend to fly away from the circle, ending up in the end of the tube.

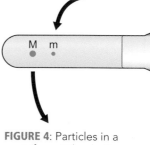

FIGURE 4: Particles in a centrifuge. Why do they move apart as the centrifuge is spun?

> ### QUESTIONS
>
> **7** Discuss what provides the centripetal force for clothes in a spin dryer.

Circular motion in action

Saturn's moons

Scientists keep discovering more moons orbiting the planet Saturn – it has at least 60 moons – yet Earth has only one. Why do planets have moons at all – and why do moons stay in orbit, rather than orbiting the Sun as planets? Gravity and centripetal force help to answer these questions.

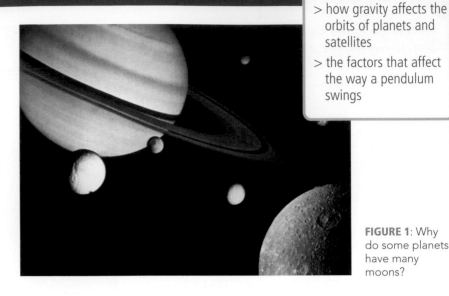

FIGURE 1: Why do some planets have many moons?

Orbits

Gravity and orbits

Gravity provides the **centripetal force** needed to keep **planets**, asteroids and comets orbiting the Sun.

The gravity between two objects increases if the mass of either object increases. That is why a large mass weighs more than a small mass. Massive planets such as Saturn often have more moons than lighter planets – the force due to gravity near them is larger, so dust and debris near them is more likely to be held in orbit.

The gravity between two objects decreases as the objects go further apart. The force due to gravity between a moon and a close planet is greater than the force between the moon and the distant Sun. This is why the Moon stays in orbit around Earth.

QUESTIONS

1 What provides the centripetal force for planets orbiting the Sun?

2 What would happen to the planets if the gravitational force stopped acting?

Satellite orbits

Satellites have many uses, such as research, monitoring and communication.

The centripetal force for a satellite is provided by gravity. Gravity is stronger nearer Earth's surface because the force of gravity is stronger as objects become closer. A satellite closer to Earth has to travel at a higher speed to prevent gravity pulling it down to Earth.

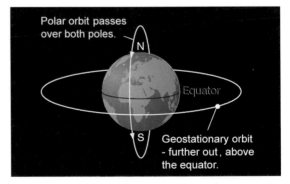

FIGURE 3: Why are weather forecast satellites put into polar orbit?

FIGURE 2: Name the eight planets in order, starting from the Sun (on the right).

A simple pendulum

A simple pendulum is a mass on a string that swings through part of a vertical circle. Although it is circular motion, it is easier to think about a pendulum in terms of energy.

At the highest points of its swing (at positions A and C in Figure 4), a pendulum stores energy in its position – **gravitational potential energy** (GPE). As the pendulum swings, this GPE is transferred to **kinetic energy** (KE) stored in the pendulum's movement. The pendulum gradually slows down because some energy is transferred to overcome friction and air resistance.

The **time period**, T, of a pendulum is the time it takes to complete one swing, from left to right and back to left again. This is related to the **frequency** of the pendulum.

$$T = \frac{1}{f}$$

where
T is periodic time in seconds (s)
f is frequency in hertz (Hz)

The longer the pendulum, the longer it takes to complete one swing and the longer its time period. For example, a 'swinging ship' ride at a theme park takes longer to complete one swing than an ordinary child's playground swing.

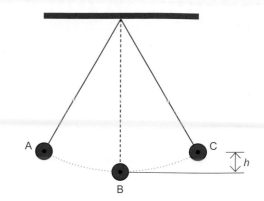

FIGURE 4: A simple pendulum.

Did you know?

Science Discovery Centres often have a long, slowly swinging pendulum to show that the Earth is turning on its own axis.

⬤ QUESTIONS

3 What happens to the energy of a pendulum as it swings?

4 A pendulum has a frequency of 4 Hz. How long will it take to complete 1 swing?

5 If you wanted a pendulum clock to slow down, would you make its pendulum longer or shorter?

Long pendulums

In Figure 5, the downward force on the pendulum bob, mass m, is its weight, mg. A component of this weight, F, makes the pendulum bob accelerate towards the centre position. For a longer pendulum, the force towards the centre is the same size, F, but the pendulum bob has further to travel. It takes longer to complete one swing.

Changing the mass, m, has no effect. Increasing m makes the force F increase, but the larger mass needs a larger force to give the same acceleration. It speeds up at exactly the same rate as a smaller mass.

⬤ QUESTIONS

6 Compare the speed of the pendulum bob at the centre position for a long pendulum and a short pendulum.

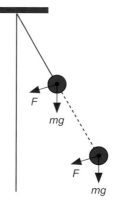

FIGURE 5: Why longer pendulums take longer for each swing.

🔍 ... long pendulum experiment ... Foucault's pendulum

Preparing for assessment: Applying your knowledge

To achieve a good grade in science, you not only have to know and understand scientific ideas, but you need to be able to apply them to other situations and investigations. These tasks will support you in developing these skills.

☀ Weapons in the Middle Ages

A trebuchet is a weapon that was in use about 500 years ago. It was used to smash defences such as city or castle walls by flinging heavy rocks at them, very fast. The trebuchet applied simple ideas about levers to fling rocks far faster than any group of soldiers could. Missiles weighing 100 kilograms could be flung 100 metres to hit a castle wall fast enough to smash the stonework.

The trebuchet consisted of a lever that could be 16 metres long. This lever was supported about one-quarter of the way along its length by a tower several metres high.

The missile, which was often a heavy rock, was placed in a pouch. The pouch was suspended using a sling a few metres below the end of the longest side of the lever. This design helped to give the missile extra speed.

A counterweight, up to 50 or 60 times heavier than the missile, was placed in a box at the other end of the lever. A team of men or horses would wind the mechanism so that the counterweight was slowly raised to its highest position. This lowered the end of the lever attached to the pouch, which would lie on the ground when the trebuchet was prepared.

When the counterweight was released, it fell to the ground. As it fell, the longer end of the lever was flung upwards. As the lever reached its highest point, the sling carried on travelling in a circle – which released the missile at its highest point.

☀ Task 1

Look at the photograph of the trebuchet.

Draw a series of sketches to show the action of a trebuchet.

☀ Task 2

Use your ideas of centre of mass to explain the following.

(a) Where will the counterweight rest after the trebuchet is fired?

(b) How does its design make the trebuchet stable during loading and firing?

☀ Task 3

When the lever reaches its highest position, the trebuchet's lever stops moving, but the sling continues to move. The missile feels a centripetal force as the sling swings above the top of the trebuchet.

(a) What causes the centripetal force and which way is it acting?

(b) How would these changes affect the size of the centripetal force felt by the missile:

(i) increasing the length of the sling

(ii) increasing the mass of the missile?

☀ Task 4

The trebuchet is loaded with a counterweight that weighs 50 000 N and a missile that weighs 1000 N. The counterweight is 4 m from the fulcrum and the missile is 12 m from the fulcrum.

(a) Calculate the moment of the missile and the moment of the counterweight.

(b) Calculate the moment needed to balance the trebuchet.

(c) Calculate the force needed to balance the trebuchet.

☀ Maximise your grade

	Answer includes showing that you...
E	recall the meaning of 'centre of mass'.
	know that a moment is the turning effect of a force.
	recall that objects travelling in a circle change direction.
	recall that increases in mass and decreases in radius will increase the centripetal force.
C	can explain that levers increase forces.
	can calculate moments using $F \times d$
	can explain that, if an object does not turn, moments are balanced.
A	can explain why objects travelling in a circle accelerate to its centre.
	can identify factors causing a centripetal force.

Checklist P3.1–3.2

To achieve your forecast grade in the exam you will need to revise

Use this checklist to see what you can do now. Refer back to the relevant topics in this book if you are not sure. Look across the three columns to see how you can progress. Bold text means Higher tier only.

Remember that you will need to be able to use these ideas in various ways, such as:

> interpreting pictures, diagrams and graphs
> suggesting some benefits and risks to society
> applying ideas to new situations
> drawing conclusions from evidence you are given.
> explaining ethical implications

Look at pages 188–209 for more information about exams and how you will be assessed.

To aim for a grade E	To aim for a grade C	To aim for a grade A
Recall that X-rays are part of the electromagnetic spectrum.	Recall precautions taken when using X-rays and CT scans.	Recall that X-rays are ionising. Explain some precautions taken when using X-rays and CT scans.
Recall that ultrasound waves are sound waves with a frequency higher than humans can hear.	Recall how ultrasound waves are produced. Recall that ultrasound waves are partially reflected from boundaries. Interpret diagrams of oscilloscope traces to obtain data.	Recall that ultrasound is not ionising. Explain how an image is built up using ultrasound waves. Use data from oscilloscope traces to calculate distances between boundaries of different materials.
Recall that ultrasound and X-rays have medical uses.	Recall properties of X-rays and ultrasound and give examples of their use in medicine, including CT scans.	Explain and evaluate the use of X-rays and CT scans for medical reasons.
Recall that a lens forms an image. Recall that a lens refracts light.	Explain the meaning of the term 'refraction'. Calculate the power of a lens. Know that the shape and refractive index of a lens affects its focal length.	Calculate refractive index. **Explain how changing the refractive index affects the shape and focal length of a lens.**
Describe the images formed by lenses. Know that lenses can magnify images.	Interpret ray diagrams for convex and concave lenses. Know that parallel rays of light are focused at the principal focus of a lens. Calculate the magnification of a lens.	Construct and interpret ray diagrams for convex and concave lenses.

To aim for a grade E

Label a diagram of the eye.

Recall that lenses can correct vision defects.

Recall that the eye and a camera have similarities and differences.

Recall that lasers are an energy source.

Recall that the mass of an object is thought to be concentrated at its centre of mass.

Recall the position of the centre of mass for a symmetrical object.

Know that the time period of a pendulum depends on the length of the pendulum.

Know that moment is the turning effect of a force.

Recall that liquids are incompressible.

Recall that objects travelling in a circle change direction but not speed.

To aim for a grade C

Describe the function of different parts of the eye.

Recall which types of lens correct different vision problems.

Compare the structure of the eye with a camera.

Recall that total internal reflection means that light can be sent along optical fibres.

Know that the centre of mass hangs below where the mass is suspended.

Describe how to find the centre of mass of a sheet of card.

Calculate the time period of a pendulum using $T = \dfrac{1}{f}$

Calculate moments using $F \times d$

Understand that, if an object does not turn, moments are balanced.

Understand that levers may be used to increase forces.

Recall that pressure in a liquid is transmitted in all directions.

Calculate pressure in hydraulic systems using force ÷ area.

Recall that increases in mass and speed, and decreases in radius will increase the centripetal force.

To aim for a grade A

Evaluate the use of lenses to correct vision defects.

Explain how lenses correct vision defects.

Calculate refractive index using $\dfrac{1}{\sin c}$ where c is the critical angle.

Calculate the size of force or distance from pivot for a balanced object.

Explain that an object topples if its weight acts outside its base.

Describe how hydraulic systems are used as force multipliers.

Recall that an object travelling in a circle accelerates to the centre because the centripetal force acts towards the centre.

Identify the force providing a centripetal force in different situations.

In the examination, equations will be given on a separate equation sheet.
Write down the equation that you will use. Show clearly how you work out your answer.

1. Doctors use ultrasound waves and X-rays to investigate medical conditions.

AO1 **(a)** What is meant by the term 'ultrasound'? [1]

AO2 **(b)** Write down one difference between X-rays and ultrasound. [1]

AO3 **(c)** Explain whether a doctor should use ultrasound or X-rays to scan an unborn baby. [3]

2. The diagram shows a cross section of the eye.

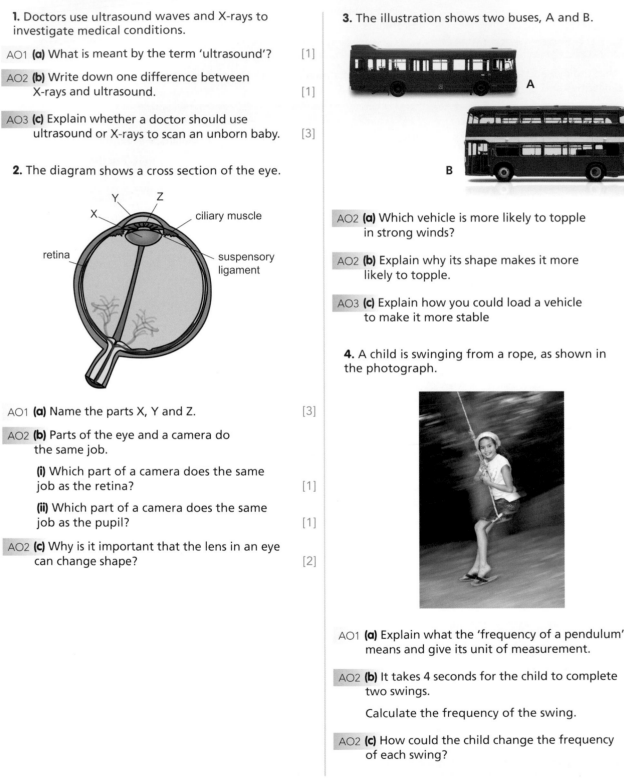

AO1 **(a)** Name the parts X, Y and Z. [3]

AO2 **(b)** Parts of the eye and a camera do the same job.

  (i) Which part of a camera does the same job as the retina? [1]

  (ii) Which part of a camera does the same job as the pupil? [1]

AO2 **(c)** Why is it important that the lens in an eye can change shape? [2]

3. The illustration shows two buses, A and B.

AO2 **(a)** Which vehicle is more likely to topple in strong winds? [1]

AO2 **(b)** Explain why its shape makes it more likely to topple. [2]

AO3 **(c)** Explain how you could load a vehicle to make it more stable [3]

4. A child is swinging from a rope, as shown in the photograph.

AO1 **(a)** Explain what the 'frequency of a pendulum' means and give its unit of measurement. [2]

AO2 **(b)** It takes 4 seconds for the child to complete two swings.

Calculate the frequency of the swing. [2]

AO2 **(c)** How could the child change the frequency of each swing? [2]

AO1 recall the science   AO2 apply your knowledge   AO3 evaluate and analyse the evidence

✳ WORKED EXAMPLE – Foundation tier

The ray diagram shows how light travels through a lens to form an image of an object.

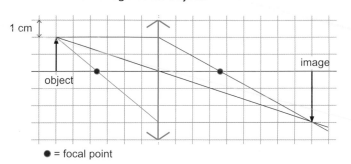

● = focal point

(a) Explain how you could use the diagrams to find the focal length of this lens. [2]

It is the distance between both principal focus points.

(b) Describe the image. [3]

It is upside down and bigger.

(c) Calculate the magnification of this lens. [3]

$$magnification = \frac{image\ height}{object\ height} = \frac{2}{3} = 0.66$$

(d) A pupil carries out an experiment using the same lens. She moves the object so that it is closer to the lens.

How does the image change when the distance between the object and lens is less than one focal length? [3]

The object is moved closer to the lens. The image becomes much bigger and is the right way up. It can't be seen on the screen but only if you look through the lens.

(e) Describe how this lens could be used as a magnifying glass. [3]

If you look through the lens, you can see the object and it is bigger.

How to raise your grade!

Take note of these comments – they will help you to raise your grade.

↓

The candidate receives only one mark.

The focal length is the distance from the centre of the lens to either of the principal focus points.

Check the marks available. This question has three possible marks. The candidate describes only two features and so receives two out of three possible marks. The candidate needed to describe the image as 'real'.

The candidate wrote the numbers down in the wrong order but still gains one mark for some correct working.

The correct answer is 3 ÷ 2 = 1.5. Remember that magnification has no units.

The candidate receives all three marks. It would have been better to use scientific language, for example, 'magnified', 'upright' and 'virtual'.

There is no need to repeat the question.

The candidate gains only two marks. The answer explains how the lens is used and what is seen, but did not say that the distance between the lens and the object must be closer than the focal length.

Be careful not to muddle up the terms 'object' and 'image'. The object did not change, but its image did.

In the examination, equations will be given on a separate equation sheet.
Write down the equation that you will use. Show clearly how you work out your answer.

1. Doctors use different methods to investigate medical problems.

AO2 **(a)** Explain why ultrasound waves are used to remove kidney stones. [2]

AO2 **(b)** Radiologists take precautions when working with X-rays.

Describe two of these precautions and explain why they are important. [4]

AO3 **(c)** Explain whether the same precautions are necessary when the radiologist is performing ultrasound scans. [3]

2. An optometrist carries out tests to choose the most suitable lens to correct a patient's vision.

AO2 **(a)** The patient's prescription shows the power of one lens as +4 dioptre.

What is the focal length of the lens? [2]

AO1 **(b)** How does the shape of the lens affect its focal length? [2]

AO3 **(c)** A lens-maker is preparing two lenses for a pair of spectacles. It is best if both lenses are a similar shape.

Explain how the lens-maker can make lenses of different strengths the same shape. [3]

3. The photograph shows a disc brake for a car wheel. It is operated by a system of hydraulics. Brake fluid in the hydraulic system transmits the force from the brake pedal to the brake discs.

AO3 **(a)** Explain two properties that brake fluid should have, to work properly in all weathers. [4]

AO2 **(b)** A car driver presses on the brake pedal with a force of 10 N. The area of the brake pedal piston is 0.001 m². The area of each of the eight pistons acting on the brake discs is 0.005 m².

Calculate the braking force acting on the car. [4]

4. The owners of a café found that the some water jugs were falling over easily when they were knocked. Other water jugs that were a different shape seemed to be more stable. All jugs held the same amount of water.

AO1 **(a)** Describe two features of a water jug that would make it more stable. [2]

AO2 **(b)** Explain why the stability of a water jug changes as it is filled with water. [3]

In part (c) of this question you will be assessed on using good English, organising information clearly and using specialist terms where appropriate.

AO2 **(c)** A new sign outside the café is going to be suspended on a single wire. The text on the sign will have to be horizontal, to be readable.

Explain how the designer could find the centre of mass of a sign formed from sheet metal and how this would help to ensure that the text is horizontal. [4]

AO1 recall the science   AO2 apply your knowledge   AO3 evaluate and analyse the evidence

✳ WORKED EXAMPLE – Higher tier

The diagram shows a racing track for motor cars.

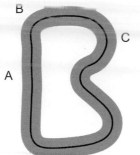

(a) One driver managed to travel around the track at a constant speed.

Explain why the vehicle had to accelerate in some places, to maintain a constant speed. [1]

The car changed direction.

> This answer is correct and receives the mark.
>
> Remember that acceleration is change of velocity (which is speed and direction).

(b) The car feels a centripetal force at point B.

What causes this centripetal force? [2]

The road

> This answer is not enough to gain a mark. The candidate should explain that friction between the tyres and road surface caused the force.

(c) Is the centripetal force strongest at point A, B or C? Explain your answer. [3]

It is stronger at B

> This is correct, but the candidate also needs to explain why.
>
> Centripetal force increases if the radius of the circle is smaller and the track bends more sharply at B.
>
> The answer receives one mark.

(d) One driver has two passengers in his car.

Explain whether this means that the driver should make adjustments to his speed as he completes a circuit. [4]

Yes he should because the car is heavier and centripetal force increases with the mass.

> The answer is correct, but the candidate did not give enough detail.
>
> The driver needs to slow down more at the bends, and not on the straight sections of the track.
>
> This answer receives two marks.

(e) When the driver stops, he applies a force of 15 N to the handle of the handbrake. Calculate the moment caused if this force is applied 25 cm from the pivot of the handbrake's handle. [2]

moment = force x distance
*   = 375 Nm*

> The candidate used the equation correctly, but forgot to change centimetres to metres and therefore loses one mark.

Physics P3.3

What you should know

Magnetism

All magnets have a magnetic field surrounding them. This magnetic field has a shape and direction.

Magnetic materials can be magnetised. Domains inside a magnetic material are arranged in the same direction when material is magnetised. They are arranged in different directions when the material is demagnetised.

Earth has a magnetic field.

 Draw a picture of the magnetic field surrounding a magnet.

Electromagnetism

An electromagnet is created when a current flows through a wire. The electromagnet is stronger if the current is larger, or if the wire is wrapped round an iron core. The electromagnet turns off when the current turns off.

Electromagnets are used in electric motors and generators.

 What happens if you change the direction of the current flowing in an electromagnet?

Electricity

Electric circuits transfer energy.

The potential difference in a circuit is a measure of the energy transferred.

Electric current can be alternating or direct.

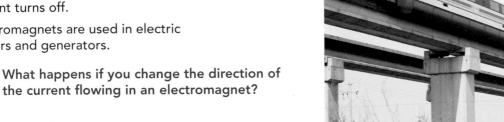

Write down the frequency of the mains supply in the UK and explain its meaning.

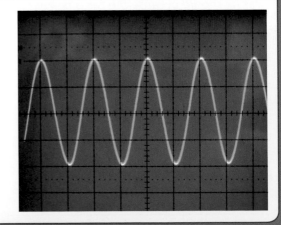

You will find out

Keeping things moving: Electromagnets

> A current flowing through a wire produces a magnetic field, creating an electromagnet.

> Electromagnets have many uses, from making doorbells ring to lifting heavy iron and steel objects.

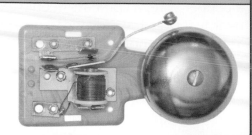

Keeping things moving: The motor effect

> Forces produced in magnetic fields can make things move. This force is called the motor effect.

> The motor effect is used in motors and other devices.

> The size and direction of the motor effect depends on several factors.

Keeping things moving: Transformers

> A potential difference is created in a wire or conductor when the wire moves near a magnet.

> Transformers consist of two coils of wire connected by an iron core. They need alternating current to work.

> Transformers change the size of potential difference in a circuit. The ratio of turns in the coils affects the potential difference ratio.

> If a transformer is 100% efficient, the power input is the same as the power output.

Keeping things moving: Switch mode transformers

> Switch mode transformers work using very high frequencies.

> Switch mode transformers are lighter and smaller than traditional transformers. They are very efficient.

> One use of switch mode transformers is in mobile phone chargers.

Electromagnetic force

You will find out:

> electric currents produce magnetic fields

> forces produced in magnetic fields can be used to make things move

The electric motor

When electricity passes through a copper wire, a compass needle nearby moves towards the wire. Yet, copper is not magnetic. Hans Oersted noticed this, in 1819. He deduced that, when an electrical current flows through a wire, it creates a magnetic force field. Electric motors use this connection between electricity and magnetism.

FIGURE 1: Electricity and magnetism create the turning force for an electric drill.

Electromagnetic force

Electromagnets

An electrical current flowing through a straight wire generates a **magnetic field**. This field acts in concentric circles around the wire. The magnetic field exerts a force on any magnetic object in the field.

A current-carrying wire wound into a long coil is called a **solenoid**. The magnetic field generated by a solenoid is the same shape as the magnetic field for a bar magnet. The coil is an **electromagnet**.

Electromagnets are useful because their magnetic field can be switched on and off by switching the current on and off. For example, a crane can use an electromagnet to pick up iron or steel, move it, and then drop it again. Electromagnets are also used in loudspeakers, electric bells and buzzers, electromagnetic door locks and a type of switch called a relay.

Using the motor effect

In Figure 3, when the current is switched on, the magnetic fields due to the wire and the horseshoe magnet interact. There is a force on the wire. A freely suspended wire will move. This is the **motor effect**.

The direction of the force, and the direction of the movement of the wire, can be reversed by:

> reversing the direction of the current (swapping the connections to the battery)

> reversing the direction of the magnetic field from the magnet (turning the magnet over so that N and S poles swap position).

If the wire is placed so that the current flows parallel to the magnetic field, there is no force on the wire.

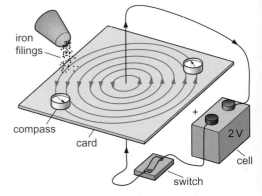

FIGURE 2: The magnetic field around a current carrying wire.

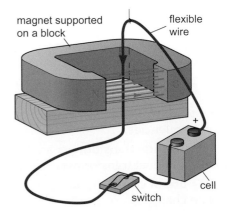

FIGURE 3: Movement of a current-carrying conductor in a magnetic field.

QUESTIONS

1 Describe what happens when a current flows through a copper wire.

2 List three devices that use an electromagnet.

3 Look at Figure 3 on the previous page.

(i) Why does the wire move in the opposite direction when you change the direction of the current?

(ii) When is there no force on the current-carrying wire?

> **Remember**
>
> A conductor will not experience a force if it is parallel to the magnetic field.

Building an electric motor

A simple electric motor is made using a coil of wire carrying a current in a magnetic field.

> There is a force on the wire in each side of the coil.

> There is no force at the ends of the coil because there the current is parallel to the magnetic field.

Fleming's left-hand rule (Figure 5) shows that the force is in opposite directions for opposite sides of the coil. Each side of the coil moves in the direction of the force, making the coil rotate.

Rotating faster

An electric motor will turn faster if the force on each side is larger. The size of the force on each wire in the magnetic field can be increased by:

> increasing the strength of the magnetic field (using a stronger magnet)

> increasing the current flowing in the wire.

The motor will also turn faster if there are more turns of wire in the coil. The total force on each side of the coil is the sum of the forces on each individual wire. A coil with more turns of wire gives a larger total force. This makes the motor turn faster.

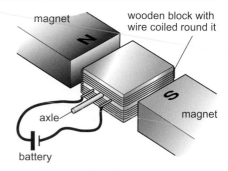

FIGURE 4: Simple electric motor.

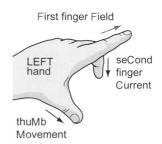

FIGURE 5: Fleming's left-hand rule.

QUESTIONS

4 Use Fleming's left-hand rule to work out which side of the coil in Figure 4 will move upwards.

5 List two ways of increasing the strength of the motor effect.

6 Explain why a motor with more turns of wire will turn faster.

> **Did you know?**
>
> Nanotechnology is discovering new ways to make electric motors that are too small to see.

Wire length

Some of the energy carried by an electric current is wasted. Wires become warm. As the current increases, the amount of energy that heats the wires increases.

If the length of wire in a motor coil is increased, there is more wire becoming warm. In a coil, there is less opportunity for this heat to transfer to the surroundings. It is more likely that the coil will become warm.

QUESTIONS

7 Suggest why commercial motors often use several coils of wire, instead of one large coil.

Electromagnetic induction

You will find out:
> about electromagnetic induction
> factors that affect the size of the induced current and potential difference

The generator effect

Electric motors and many other devices need electricity to work. Generating electricity is crucial to health and wellbeing in the 21st century. The world has come a long way since Michael Faraday, in 1831, worked out how to use a moving wire and a magnetic field to generate electricity.

FIGURE 1: Enough electricity can be generated on the wheel of a bicycle to power a lamp.

Generating electricity

Michael Faraday discovered that, when a metal wire (a conductor)
1. is moved through the field lines of a magnet and
2. is in a complete circuit,
an electric current flows in the wire.

A sensitive ammeter or galvanometer in the circuit will show the current flowing. The current flows in a direction that opposes the movement.

Faraday said that a **potential difference** is induced across the ends of the wire that 'pushes' the electrons, making the current. This is **electromagnetic induction**.

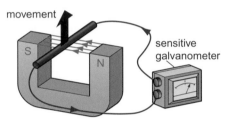

FIGURE 2: Moving a wire through a magnetic field. What happens if the wire moves in the opposite direction?

Coils and magnets

Figure 3 shows how a moving magnet induces a potential difference across a coil of wire. This generates an electric current. Keeping the magnet still and moving the coil also induces a potential difference. If the induced potential difference is large, the current that flows will be large.

Faraday showed that:

> current flows only when the magnet is moving

> current increases when

  (a) using a stronger magnet

  (b) moving the magnet or the coil faster

  (c) increasing the number of turns of wire in the coil

  (d) increasing the diameter of the coil.

Did you know?

Generators are sometimes called alternators. In a car, the alternator uses the car's movement to generate the electricity that keeps the battery charged up.

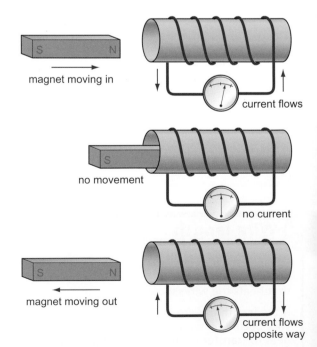

FIGURE 3: A current flows only when the magnet is moving.

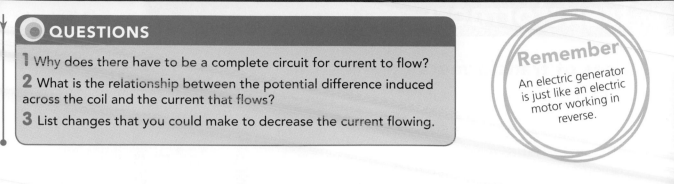

Remember

An electric generator is just like an electric motor working in reverse.

QUESTIONS

1 Why does there have to be a complete circuit for current to flow?

2 What is the relationship between the potential difference induced across the coil and the current that flows?

3 List changes that you could make to decrease the current flowing.

Coils for magnets

Faraday did more experiments to find out how to induce (cause) a current to flow in a coil of wire. He replaced the moving magnet with a coil of wire. He found that:

> a constant current in the left-hand coil of wire did not induce a current in the other coil

> when the current in the left-hand coil was changing, a current was induced in the other coil

> when the current in the left-hand coil is constant, the magnetic field around the coil is constant.

Comparing Figure 4 with Figure 3, it shows that:

> when the magnet or coil is not moving, no current is induced

> when the current in the left-hand coil changes it is the same as a magnet moving and a current is induced

> changing the current is like moving the magnet faster, so the induced current is larger.

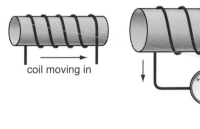

coil moving in · current flows

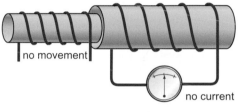

no movement · no current

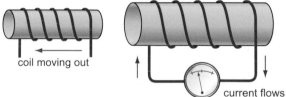

coil moving out · current flows opposite way

FIGURE 4: Inducing a current.

QUESTIONS

4 State the conditions for there to be an induced potential difference and a current flowing.

5 Explain in your own words why an alternating current in the left-hand coil induces a current in the other coil, but a direct current in the left-hand coil does not.

Surge protection

Sensitive equipment, such as computers, often has built-in surge protection.

When the electricity supply to equipment is switched on or switched off, the size of the current changes very quickly as current begins to flow or stops flowing. This changing current can induce a large current through components in sensitive equipment, causing damage.

Surge protection diverts 'extra' current away from delicate components.

QUESTIONS

6 Research and then briefly describe how surge protection works.

Transformers

You will find out:
> about the structure of transformers
> how transformers work

Uses of transformers

In the UK, a.c. mains electricity is supplied at 230 V. Many appliances need a much smaller potential difference (voltage) than this. A door bell, your computer or your sound system have small components that need only 12 V, or even less. Transformers reduce – step down – mains potential difference to a smaller value.

FIGURE 1: The potential difference across electronic components has to be much less than 230 V.

Step-up and step-down transformers

Transformers are devices that either increase or decrease potential difference (p.d.). When a p.d. is connected across one coil, the **primary coil**, this causes a p.d. across the output of the other coil, the **secondary coil**.

A **step-up transformer** increases the p.d.

> The p.d. across the secondary coil is greater than the p.d. across the primary coil.

> The secondary coil has more turns of wire than the primary coil.

A **step-down transformer** decreases the p.d.

> The p.d. across the secondary coil is less than the p.d. across the primary coil.

> The secondary coil has less turns of wire than the primary coil.

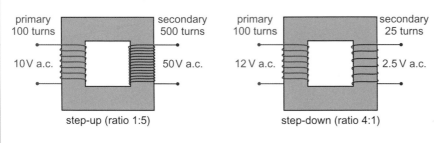

FIGURE 2: Step-up and step-down transformers.

Did you know?

Transformers vary in size from the massive ones in the National Grid to the tiny ones in toys.

QUESTIONS

1 In a transformer, what is
(a) the primary coil
(b) the secondary coil?

2 How would you recognise the difference between a step-up and step-down transformer?

3 A computer that plugs into the mains needs to operate on a p.d. of 12 V. Explain whether you would use a step-up or a step-down transformer.

How transformers work

Transformers have two coils of wire wound around a laminated, **soft iron core**.

> The two coils are close to each other but not touching.

> The coils are insulated from each other and from the iron core.

> Current does not flow in the core.

> The wire used is insulated so that the current does not short circuit between coils.

Q ... step up AND step down transformers potential difference

When an alternating potential difference is put across the primary coil, an alternating current (a.c.) flows in the primary coil.

The soft iron core is magnetic. The alternating current sets up an alternating magnetic field in the soft iron core.

This alternating magnetic field causes an alternating current in the secondary coil, and an alternating potential difference across the secondary coil.

Transformer calculations

The potential difference across the secondary coil of a transformer depends on the potential difference across the primary coil and on the number of turns of wire on each coil. An equation connects these four values.

$$\frac{V_p}{V_s} = \frac{n_p}{n_s}$$

where

V_p is the potential difference across the primary coil in volts (V)

V_s is the potential difference across the secondary coil in volts (V)

n_p is the number of turns on the primary coil

n_s is the number of turns on the secondary coil

Example

A step-down transformer has an output p.d. across the secondary coil of 12 V. There are 1000 turns of wire on the primary coil and 50 turns of wire on the secondary coil. What is the p.d. across the input of the primary coil?

$$\frac{V_p}{12} = \frac{1000}{50}$$

Therefore, V_p = 12 × 20 = 240 V

The potential difference across the input of the primary coil is 240 V.

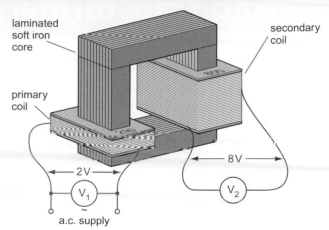

FIGURE 3: The coils in the transformer do not touch. What else do you notice about the coils?

● QUESTIONS

4 Describe how there is a potential difference across the secondary coil of a transformer, when the two coils are not connected to each other.

5 "A transformer works with a.c. or d.c. current." Is this true or false? Explain your answer.

6 A transformer has 150 turns of wire on the primary coil and 600 turns of wire on the secondary coil.

(i) Is it a step-up or a step-down transformer?

(ii) If the output p.d. across the secondary coil is 800 V, what is the input p.d. across the primary coil?

Transformer size

The ratio of the p.d. across the primary coil to the p.d. across the secondary coil ($V_p : V_s$) is always exactly the same as the ratio of the number of turns on the primary coil to the number of turns on the secondary coil ($n_p : n_s$).

These ratios can be used to calculate missing values of p.d or numbers of turns.

● QUESTIONS

7 A transformer is required to change an input p.d. of 25 000 V to an output p.d. of 40 000 V.

(i) Use ratios to suggest several possible pairs of values for the number of turns on primary and secondary coils.

(ii) If the secondary coil had 1600 turns, how many turns would the primary coil have?

Q ... electromagnetism transformer equations

Using transformers

You will find out:
> about the efficiency of transformers
> how switch mode transformers work and their advantages

Danger, overhead cables

The potential difference across power cables that transmit electricity over long distances may be up to 400 kV. Transformers are used to step up the potential difference to these high values, at the generating station. 230 V is supplied to our homes through a series of step-down transformers.

FIGURE 1: Do you know why overhead power cables are so dangerous?

DANGER
OVERHEAD ELECTRIC POWER LINES
NO FISHING BEYOND THIS POINT

Transformers in the National Grid

The **National Grid** transfers electricity from power stations, where it is generated, to homes and businesses, where it is used. As the current flows through the **power transmission cables**, some of the energy is transferred to the surroundings as heat. The energy 'wasted' in this way can be reduced by transmitting the energy with very high potential difference.

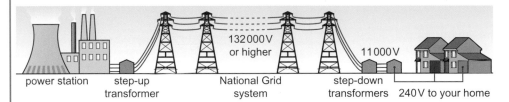

132 000 V or higher

11 000 V

power station | step-up transformer | National Grid system | step-down transformers | 240 V to your home

FIGURE 2: Step-up and step-down transformers in the National Grid.

> Step-up transformers at the power station increase the p.d. output from the power station to the very high values used in the transmission cables.

> In towns or cities, an area sub-station uses step-down transformers to decrease the p.d. before the electrical energy is transmitted to local areas.

> Smaller, local sub-stations have more step-down transformers that decrease the p.d. again, down to the 230 V used in our homes.

These small, fenced-off sub-stations have safety signs that warn of the danger of electrocution.

QUESTIONS

1 Explain in your own words where transformers are used in the National Grid.

2 Explain in your own words why the National Grid uses such high values of potential difference.

FIGURE 3: Local sub-stations contain step-down transformers, with an output at 230 V.

Transformers in use

Transformers and efficiency

If transformers were 100% efficient, all of the electrical energy going into the transformer each second would be transferred out again. The **electrical power** output of the transformer would equal the electrical power input.

For a transformer that is 100% efficient:

$$V_p \times I_p = V_s \times I_s$$

where

V_p is the potential difference across the primary coil in volts (V)
I_p is the current in the primary coil in amperes (amps, A)
V_s is the potential difference across the secondary coil in volts (V)
I_s is the current in the secondary coil in amperes (amps, A)

In practice, transformers are not 100% efficient. Transformers (often called 'mains adaptors' or – incorrectly – 'power supplies') for appliances such as laptop computers become warm when they are used. Some of the energy going into the transformer is transferred to the surroundings by heating.

Switch mode transformers

Switch mode transformers are small, light, modern transformers such as those used in mobile phone chargers. The charger could be even smaller, except that it must fit a three-pin mains socket.

Switch mode transformers use a much higher frequency a.c. current than traditional transformers – usually between 50 kHz and 200 kHz, instead of the traditional 50 Hz mains supply.

If nothing is connected to the transformer output when they are plugged in, switch mode transformers do not use much power. This is because they are small.

FIGURE 4: Switch mode transformers in mobile phone chargers can be very small.

QUESTIONS

3 The primary coil of a transformer has a p.d of 240 V, and a current of 20 mA. If the p.d across the secondary coil is 12 V, calculate the current in this coil.

4 Name one factor that limits how small a switch mode transformer for a mobile phone charger can be made.

5 Explain what is meant by an 'a.c. current with a frequency of 50 kHz'.

6 Suggest some advantages of a switch mode transformer, compared with a more traditional transformer.

How switch mode transformers work

When a magnet moves towards a coil of wire, the changing magnetic field causes a current in the coil.

The a.c. current in the primary coil of a transformer is continually changing direction. This produces a continually changing magnetic field, rather like continually moving a magnet towards and away from the secondary coil.

If the a.c. current changes more quickly – with a higher frequency – it is like moving the magnet more quickly. A weaker magnet can be used to give the same size current in the secondary coil.

This means that a smaller magnetic core can be used.

QUESTIONS

7 Explain in your own words the link between the frequency of the current in a transformer and the magnetic field in the core.

Preparing for assessment: Analysing and interpreting data

To achieve a good grade in science, you not only have to know and understand scientific ideas, but you need to be able to apply them to other situations and investigations. These tasks will support you in developing these skills.

✺ Investigating transformers

A potential difference is induced across the ends of a coil of wire when it is exposed to a changing magnetic field. This effect is used in transformers to increase or decrease a.c. potential differences.

Some students carried out an investigation to find out if the *turns rule for a transformer* always applies.

> The students used an a.c. supply.

> They adjusted the supply to set the potential difference across the primary coil for each set of readings.

Results

The table shows the students' results.

Number of turns on the primary coil	Number of turns on the secondary coil	Ratio turns on primary: turns on secondary	Potential difference across primary coil	Potential difference across secondary coil	Ratio potential difference across primary: potential difference across secondary
120	60		2.0	1.01	
120	120		2.0	1.95	
120	240		2.0	3.95	
240	60		2.0	0.50	
240	120		2.0	1.02	
240	240		2.0	1.90	
60	60		2.0	1.90	
60	120		2.0	3.95	
60	240		2.0	7.80	

✺ Processing data

1. Copy the results table. Complete the shaded columns to show the ratio of turns and the ratio of potential differences.

> Tables should always include space for all readings (raw data). The data can be processed later and included in other columns.

✳ Analysing data

2. Describe the pattern for the ratios that you have calculated.

> Describe how the dependent variable changes when the independent variable changes.

3. The students felt that these results were enough to prove that the turns rule applied whenever transformers were used. It usually good practice to repeat readings in an experiment.

Explain why. Which values would you repeat?

> It can be worth repeating an experiment using different values of a controlled variable, or over a different range.

4. A higher input potential difference can cause output potential differences that are too dangerous for use in a school laboratory. The students had to limit the values of input potential differences that they used.

Explain whether this limited their conclusions.

> Remember that patterns may change when a different range of readings is taken.

✳ Interpreting data

5. The students could have repeated each reading several times, but chose not to.

Explain whether this means that their data were unreliable.

> If an experiment is reliable, several different sets of data will show the same relationship.

6. The voltmeters that the students used measured potential differences to one decimal place.

Explain whether the students would have improved their results by using more precise data loggers to record the potential differences.

> Think what the students were asked to find out and whether a data logger would improve their final conclusion.

7. Suggest some improvements to the experiment.

> Improvements should reduce the chance of errors or improve the quality of results.

✳ Connections

How Science Works
- Collect primary and secondary data
- Select and process primary and secondary data
- Analyse and interpret primary and secondary data
- Use scientific models and evidence to develop hypotheses, arguments and explanations

Science ideas
3.3.2 Transformers

Checklist P3.3

To achieve your forecast grade in the exam you will need to revise

Use this checklist to see what you can do now. Refer back to the relevant topics in this book if you are not sure. Look across the three columns to see how you can progress.

Remember that you will need to be able to use these ideas in various ways, such as:

> interpreting pictures, diagrams and graphs

> applying ideas to new situations

> explaining ethical implications

> suggesting some benefits and risks to society

> drawing conclusions from evidence you are given.

Look at pages 188–209 for more information about exams and how you will be assessed.

To aim for a grade E	To aim for a grade C	To aim for a grade A
Recall that current flowing through a wire creates a magnetic field.	Recall that the motor effect occurs when forces in magnetic fields make things move. Recall how to change the size or direction of the motor effect. Use Fleming's left-hand rule.	
Recall some uses of electromagnets.	Describe uses of the motor effect.	
Recall that moving a magnet into a wire coil induces a current.	Describe the current when a magnet is moved into a coil of wire.	Explain why a potential difference is induced in terms of field lines.
Name the primary coil, the secondary coil and the soft iron core in a transformer.	Describe the structure of a transformer.	
Know that a transformer needs alternating current (a.c.) potential difference to work.	Know that a transformer has a changing magnetic field inside the core.	Describe how a transformer works using alternating current (a.c.) potential difference.

To aim for a grade E   To aim for a grade C   To aim for a grade A

Recall that a transformer changes the size of potential difference.

Recall that a step-up transformer increases the potential difference and a step-down transformer decreases the potential difference across the ends of the secondary coil.

Use
$$\frac{V_p}{V_s} = \frac{N_p}{N_s}$$
to calculate the change in potential difference or the number of turns in the coils.

Know that transformers have a power input and a power output.

Know that power output = power input in a 100% efficient transformer.

Use
$$V_p \times I_p = V_s \times I_s$$
to calculate the potential difference across or the current in the coils.

Know that switch mode transformers are lighter and smaller than traditional transformers.

Know that switch mode transformers work at frequencies between 50 kHz and 200 kHz.

Know that switch mode transformers use very little power when switched on with no load.

Recall uses of switch mode transformers.

Describe applications of switch mode transformers.

Compare the use of switch mode transformers with traditional transformers, for different applications.

In the examination, equations will be given on a separate equation sheet.
Write down the equation that you will use. Show clearly how you work out your answer.

1. The diagram shows a transformer that a student is using in an experiment.

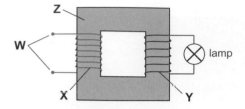

AO1 **(a)** Label W, X, Y and Z . Use words from the list below. [4]

- a.c. supply
- iron core
- primary coil
- secondary coil

AO2 **(b)** Explain why the transformer does not work when the student uses a battery as a power supply. [2]

AO3 **(c)** The student is making another transformer using two sets of coils. One set has 60 turns and the other set has 120 turns. If he wants the transformer to increase the voltage, explain which coil should be used as the primary coil. [3]

AO1 **(d)** Give the name for this type of transformer. [1]

2. A teacher connected an ammeter to a coil of wire. The ammeter reading was 0 amps.

AO1 **(a)** What happened to the reading on the ammeter when the teacher moved a magnet into the coil of wire? [1]

AO2 **(b)** He moved the magnet out of the coil of wire. What happened to the reading on the ammeter? [2]

3. Switch mode transformers are often used in mobile phone chargers.

AO1 **(a)** Write down two differences between traditional transformers and switch mode transformers. [2]

AO1 **(b)** A mobile phone charger is plugged in without the mobile phone being attached.

Explain why it uses very little power. [1]

4. The diagram shows an electric motor.

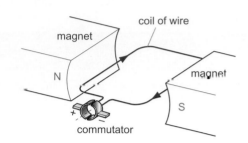

AO1 **(a)** Put the following sentences in order, to explain how the motor works: [5]

1 The magnetic field around the wire acts with the magnetic field from the magnets.

2 The coil of wire spins round.

3 One side of the coil is pushed up and one side is pushed down.

4 An electric current flows through the wire.

5 A magnetic field is produced around the wire.

AO2 **(b)** A factory uses machinery to mix very large quantities of food. The motor creates a larger force than the motor in a household food mixer.

Describe two ways in which a motor can be modified to increase the size of the force. [2]

5. A student made a motor during a lesson. The motor turned slowly when the circuit was complete.

AO1 **(a)** Give two ways that the student can make the motor spin more quickly. [2]

AO2 **(b)** How can the student make the motor spin the other way? [1]

AO1 **(c)** Name one piece of equipment that uses this effect. [1]

AO1 recall the science   AO2 apply your knowledge   AO3 evaluate and analyse the evidence

✱ WORKED EXAMPLE – Foundation tier

(a) The crane in the illustration uses an electromagnet to sort different materials.

Why is an electromagnet more suitable to move heavy pieces of metal than a permanent magnet? [2]

The electromagnet attracts iron and steel and can be turned on and off.

> Always answer questions carefully – as all magnets attract iron and steel, the candidate does not gain a mark for this point.
>
> The answer receives only one mark.
>
> The candidate should also have said that electromagnets can be very strong.

(b) The diagram below shows a loudspeaker. The coil of wire is attached to the paper cone. The coil of wire slips into a gap between the poles of a permanent magnet.

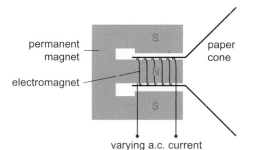

permanent magnet

paper cone

electromagnet

varying a.c. current

(i) Explain why the paper cone moves when a current flows in the coil of wire. [3]

The coil of wire becomes magnetic and reacts with the magnet.

> The candidate gains just two out of three marks.
>
> The answer would improve if it referred clearly to the permanent magnet or the electromagnet. For the third mark it must say a force is produced that makes the paper cone move.

> The candidate gains one mark.
>
> The answer should include the fact that the direction of an a.c. current keeps changing.

(ii) Why does the direction of movement of the electromagnet keep changing? [2]

The current is a.c.

(iii) Explain how you can change the a.c. current to produce a quieter sound. [2]

Make the current smaller.

> The candidate gains one out of two marks.
>
> The answer should say that a smaller current produces a smaller force (and less movement).

In the examination, equations will be given on a separate equation sheet.
Write down the equation that you will use. Show clearly how you work out your answer.

1. The diagram shows a simple motor.

AO2 **(a)** Explain why the coil of wire moves when a current flows through the coil. [3]

AO1 **(b)** Draw lines to show the direction of the magnetic field between the two magnets. [2]

AO2 **(c)** State and use Fleming's left-hand rule to calculate the direction that the coil of wire moves, when a current flows in the wire. [3]

AO2 **(d)** The current through the wire is increased. Explain how this affects the motor. [2]

2. A student attaches a magnet to a spring and suspends the spring from a clamp stand. The spring is stretched slightly and then released so that the magnet moves up and down. The student connects a coil of wire to an a.c. ammeter and places the coil of wire so that the magnet can move up and down inside it.

AO2 **(a)** Describe what the student will see on the ammeter. [3]

AO2 **(b)** What changes to the experiment can the student make, to increase the effect? [2]

3. A student carried out some experiments with transformers. The table below includes some of the student's results.

Turns on primary coil	Turns on secondary coil	Input voltage	Output voltage
120	120	6	A
120	240	6	B
120	60	C	6

AO2 **(a)** Complete the missing figures, A, B and C, in the table. [3]

AO2 **(b)** The transformer in a power station is assumed to be 100% efficient. The power station generates electricity at 25 000 V. This causes a current of 8 000 A in the wires.

Calculate the current flowing through the power lines if the voltage is stepped up to 400 000 V. [3]

In part (c) of this question you will be assessed on using good English, organising information clearly and using specialist terms where appropriate.

AO3 **(c)** A new material has been designed for possible use in transformer cores. The material is lightweight, easily magnetised and demagnetised, and does not conduct electricity.

Explain whether it is suitable for use in the core. [5]

4. Switch mode transformers are used in mobile phone chargers.

AO1 **(a)** At what frequency range do most switch mode transformers operate? [1]

AO2 **(b)** Explain two reasons why switch mode transformers are more suitable for use in mobile phone chargers than traditional transformers. [4]

AO1 **(c)** Explain why transformers only work with alternating current. [2]

AO2 **(d)** A mobile phone recharger operates at a potential difference of 6 V. Mains electricity is supplied at 230 V.

If the primary coils of the recharger have 2300 turns, how many turns are there on the secondary coil of the transformer? [3]

AO1 recall the science   AO2 apply your knowledge   AO3 evaluate and analyse the evidence

✱ WORKED EXAMPLE – Higher tier

The diagram shows a simple microphone. A coil of thin wire is wrapped around a magnet. The wire is attached to a thin sheet of material, which vibrates when sound waves reach it.

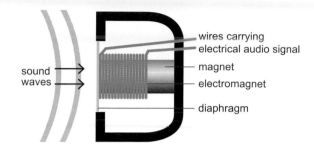

sound waves

wires carrying electrical audio signal
magnet
electromagnet
diaphragm

(a) A potential difference is created in the coil of wire when a sound wave reaches the microphone.

Explain why. [3]

The material vibrates which makes the coil vibrate.

(b) Louder sound waves have larger amplitudes.

Why does this increase the electrical signal? [2]

Louder sounds make the coil of wire move further and faster, which increases the induced voltage.

(c) A microphone converts sound waves to an alternating electrical signal. It can be converted to a loudspeaker, which converts an alternating electrical signal to sound waves.

Explain how the microphone creates sounds if it is connected to an alternating electrical signal. [5]

The electrical signal makes the coil of wire vibrate and this makes the sheet of material vibrate creating sound waves.

> **How to raise your grade!**
> Take note of these comments – they will help you to raise your grade.

> The candidate gains just two out of three marks. The answer should also say either that a potential difference is produced when the coil cuts through the magnetic field lines or when the coil moves in the magnetic field.

> The candidate gains both marks.
> Remember to link ideas about electromagnetic induction with the example in the question.

> This answer gains three out of five marks.
> The candidate should say that changing current in the coil of wire creates a changing magnetic field surrounding it. This interacts with the magnetic field surrounding the magnet, which produces a force.

Carrying out practical investigations in GCSE Separate Sciences

Introduction

As part your GCSE Separate Sciences course, you will develop practical skills and have to carry out investigative work in science.

Your investigative work will be divided into several parts:

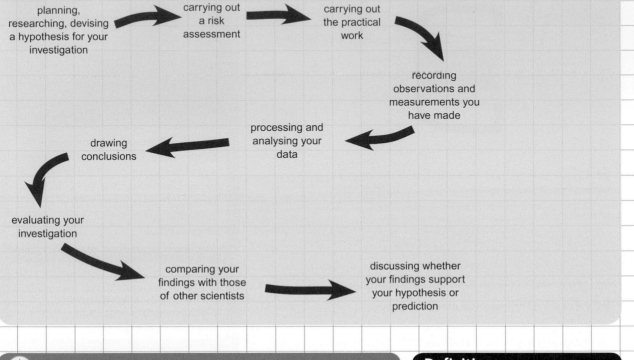

planning, researching, devising a hypothesis for your investigation → carrying out a risk assessment → carrying out the practical work → recording observations and measurements you have made → processing and analysing your data → drawing conclusions → evaluating your investigation → comparing your findings with those of other scientists → discussing whether your findings support your hypothesis or prediction

✴ Planning and researching your investigation

A scientific investigation usually begins with you thinking about an idea, answering a question, or trying to solve a problem.

Researching what other people know about the idea, question or problem may well suggest some variables that have an effect on what you decide to investigate.

From this you should develop a hypothesis. For example you might notice that plants grow faster in a heated greenhouse than an unheated greenhouse.

Your hypothesis would be that 'the rate of photosynthesis is increased by the temperature of the environment in which the plant is grown.'

You would then plan how you will carry out an investigation to test this hypothesis.

To formulate a hypothesis you are likely to need to research some of the background science.

First of all, use your lesson notes and your textbook. The topic you have been given to investigate will relate to the science that you have learned in class.

Also make use of the internet, but make sure that your internet search is closely focused on the topic that you are investigating.

Definition

A **hypothesis** is a possible explanation that someone suggests to explain some scientific observations.

Assessment tip

When devising your hypothesis, it is important that it is testable. In other words, you must be able to test the hypothesis in the school lab.

Assessment tip

You need to use your research to explain why you made your hypothesis.

✔ The search terms you use on the internet are very important. 'Investigating temperature and photosynthesis' is a better search term than just 'photosynthesis', as it is more likely to provide links to websites that are more relevant to your investigation.

✔ The information on websites also varies in its reliability. Free encyclopaedias often contain information that has not been written by experts. Some question and answer websites might appear to give you the exact answer to your question, but be aware that they may sometimes be incorrect.

✔ Most GCSE Science websites are more reliable but, if in doubt, use other information sources to verify the information.

As a result of your research, you may be able to extend your hypothesis and make a **prediction** based on science.

> ### Example 1
>
> Investigation: Plan and research an investigation into the effect of pH on how well yeast catalyses the fermentation of sugars.
>
> Your hypothesis might be 'I think that the yeast will produce the most carbon dioxide per minute at pH7'.
>
> You should be able to justify your hypothesis by some facts that you find. For example 'yeast contains enzymes and I know that most enzymes work best in neutral pH solutions'.

✳ Choosing a method and suitable apparatus

As part of your planning, you must choose a suitable way of carrying out the investigation.

You will have to choose suitable techniques, equipment and technology, if this is appropriate. How do you make this choice?

You will have already carried out the techniques you need to use during the course of practical work in class (although you may need to modify these to fit in with the context of your investigation). For most of the experimental work you do, there will be a choice of techniques available. You must select the technique:

✔ that is most appropriate to the context of your investigation, and

✔ that will enable you to collect valid data, for example if you are measuring the effects of light intensity on photosynthesis, you may decide to use an LED (light-emitting diode) at different distances from the plant, rather than a light bulb. The light bulb produces more heat, and temperature is another independent variable in photosynthesis.

Your choice of equipment, too, will be influenced by the measurements that you need to make. For example:

✔ you might use a one-mark or graduated pipette to measure out the volume of liquid for a titration, but

✔ you may use a measuring cylinder or beaker when adding a volume of acid to a reaction mixture, so that the volume of acid is in excess to that required to dissolve, for example, the calcium carbonate.

Assessment tip

Make sure that you make a detailed note of which sources you have used: for a book, the author's name and the title; for a website, the name of it.

Assessment tip

Higher tier
You are expected to be able to balance chemical equations. So, for example, if the enzyme is being used to decompose hydrogen peroxide to water and oxygen you should be able to balance the equation: $H_2O_2 \rightarrow H_2O + O_2$ to give $2H_2O_2 \rightarrow 2H_2O + O_2$

Assessment tip

Technology, such as data-logging and other measuring and monitoring techniques, for example heart sensors, may help you to carry out your experiment.

Definition

The **resolution** of the equipment refers to the smallest change in a value that can be detected using a particular technique.

Assessment tip

Carrying out a preliminary investigation, along with the necessary research, may help you to select the appropriate technique to use.

✴ Variables

In your investigation, you will work with independent and dependent variables.

The factors you choose, or are given, to investigate the effect of are called **independent variables**.

What you choose to measure, as affected by the independent variable, is called the **dependent variable**.

✴ Independent variables

In your practical work, you will be provided with an independent variable to test, or will have to choose one – or more – of these to test. Some examples are given in the table.

Investigation	Possible independent variables to test
activity of amylase enzyme	> temperature > sugar concentration
rate of a chemical reaction	> temperature > concentration of reactants
stopping distance of a moving object	> speed of the object > the surface on which it's moving

Independent variables can be **categoric** or **continuous**.

> When you are testing the effect of different disinfectants on bacteria you are looking at categoric variables.
> When you are testing the effect of a range of concentrations of the same disinfectant on the growth of bacteria you are looking at continuous variables.

Range

When working with an independent variable, you need to choose an appropriate **range** over which to investigate the variable.

You need to decide:

✔ which treatments you will test, and/or
✔ the upper and lower limits of the independent variables to investigate, if the variable is continuous.

Once you have defined the range to be tested, you also need to decide the appropriate intervals at which you will make measurements.

The range you would test depends on:

✔ the nature of the test
✔ the context in which it is given
✔ practical considerations
✔ common sense.

Definition

Variables that fall into a range of separate types are called **categoric variables**.

Definition

Variables that have a continuous range, or are numeric, are called **continuous variables**.

Definition

The **range** defines the extent of the independent variables being tested, for example, from 15 cm to 35 cm.

Example 2

1 Investigation: Investigating the energy changes that occur when different fuels are burned.

You may decide which fuels to test based on a range that you have been provided with. You may wish to test only a liquid or you may choose to test both solid and liquid fuels. You will not test certain fuels, such as pressurised gas, because of safety reasons.

2 Investigation: Comparing the focal lengths of a number of lenses.

The range of lenses you can choose will depend on the availability of the resources at your school. As a minimum though you would choose two different lenses to compare, such as convex and plane or concave and convex.

Temperature

You might be trying to find out the best temperature at which to grow tomatoes.

The 'best' temperature is dependent on a number of variables that, taken together, would produce tomatoes as quickly as possible whilst not being too costly. You might use a high temperature, but the cost of the fuel may outweigh the advantage of growing the tomatoes more quickly.

You should limit your investigation to just one variable, temperature, and keep the other factors constant (such as watering and feeding the tomatoes). Later you can consider other variables, including fuel costs for heating.

> **Assessment tip**
>
> Again, it is often best to carry out a trial run or preliminary investigation, or carry out research, to determine the range to be investigated.

✳ Dependent variables

The dependent variable may be clear from the problem that you are investigating, for example the stopping distance of moving objects. You may have to make a choice.

Example 3

1 Investigating the amount of sodium hydroxide solution needed to neutralise $25cm^3$ of hydrochloric acid.

There are several ways that you could establish the neutralisation point in this investigation. These include:

> measuring the volume of sodium hydroxide solution being added by counting the number of drops or, more precisely, by using a pipette

> identifying the neutralisation point using an indicator such as methyl orange or a pH probe to measure pH.

> **Assessment tip**
>
> The value of the *depend*ent variable is likely to *depend* on the value of the independent variable. This is a good way of remembering the definition of a dependent variable.

2 Investigation: Measuring the rate of a chemical reaction

You could measure the rate of a chemical reaction in the following ways:

> the rate of formation of a product

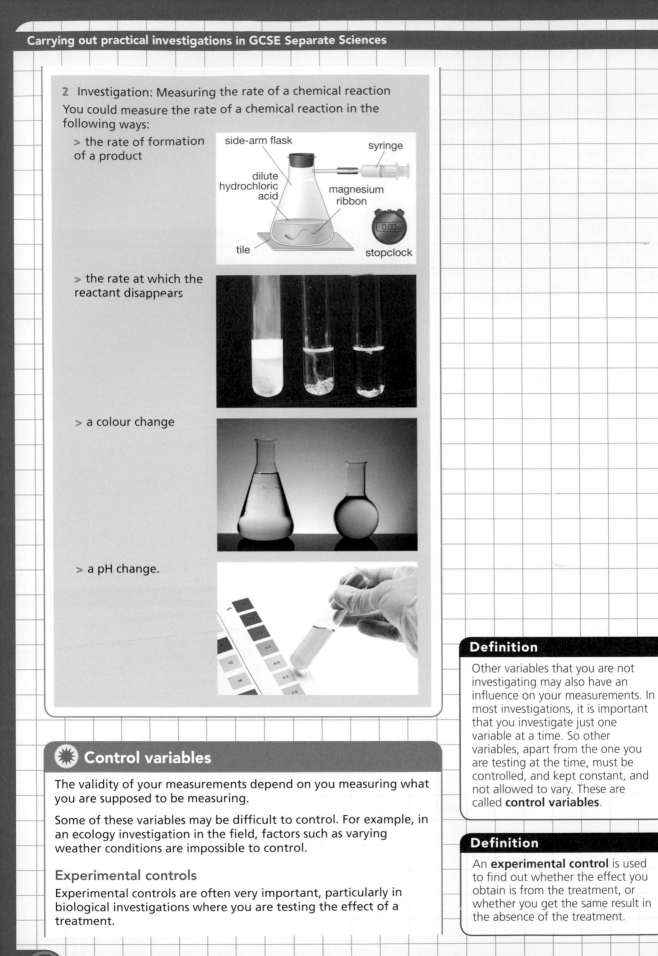

side-arm flask

syringe

dilute hydrochloric acid

magnesium ribbon

tile

stopclock

> the rate at which the reactant disappears

> a colour change

> a pH change.

Control variables

The validity of your measurements depend on you measuring what you are supposed to be measuring.

Some of these variables may be difficult to control. For example, in an ecology investigation in the field, factors such as varying weather conditions are impossible to control.

Experimental controls

Experimental controls are often very important, particularly in biological investigations where you are testing the effect of a treatment.

Definition

Other variables that you are not investigating may also have an influence on your measurements. In most investigations, it is important that you investigate just one variable at a time. So other variables, apart from the one you are testing at the time, must be controlled, and kept constant, and not allowed to vary. These are called **control variables**.

Definition

An **experimental control** is used to find out whether the effect you obtain is from the treatment, or whether you get the same result in the absence of the treatment.

Example 4

Investigation: Comparing the rate of reaction between ethanoic acid and hydrochloric acid with marble chips.

Many factors affect the rate of a chemical reaction, so you need to control those factors that you are not investigating. You are comparing ethanoic acid and hydrochloric acid. Therefore you need to make sure the volumes and concentrations are the same (for example, 25 cm^3 of 1 mol/dm^3 acid), the quantity of marble chips and their sizes are the same and the temperature is the same.

✳ Assessing and managing risk

Before you begin any practical work, you must assess and minimise the possible risks involved.

Before you carry out an investigation, you must identify the possible hazards. These can be grouped into biological hazards, chemical hazards and physical hazards.

Biological hazards include:	Chemical hazards can be grouped into:	Physical hazards include:
> microorganisms	> irritant and harmful	> equipment
> body fluids	> toxic	> objects
> animals and plants.	> oxidising	> radiation.
	> corrosive	
	> harmful to the environment.	

Scientists use an international series of symbols so that investigators can identify hazards.

Hazards pose risks to the person carrying out the investigation.

A risk posed by chlorine gas produced in the electrolysis of sodium chloride will be reduced if you devise a method to extract the gas from the process plant so that workers do not inhale it.

When you use hazardous materials, chemicals or equipment in the laboratory, you must use them in such a way as to keep the risks to absolute minimum. For example, one way is to wear eye protection when using hydrochloric acid.

Any action that you carry out to reduce the risk of a hazard happening is known as a 'control measure'.

Definition

A **hazard** is something that has the potential to cause harm. Even substances, organisms and equipment that we think of being harmless, used in the wrong way, may be hazardous.

Hazard symbols are used on chemical bottles so that hazards can be identified.

Definition

The **risk** is the likelihood of a hazard causing harm in the circumstances it's being used in.

Assessment tip

When assessing risk and suggesting control measures, these should be specific to the hazard and risk, and not general. Hydrochloric acid is dangerous as it is 'corrosive – skin and eye contact should be avoided' will be given credit, but 'wear eye protection' is too vague.

✳ Risk assessment

Before you begin an investigation, you must carry out a risk assessment. Your risk assessment must include:

✔ all relevant hazards (use the correct terms to describe each hazard, and make sure you include them all, even if you think they will pose minimal risk)

✔ risks associated with these hazards

✔ ways in which the risks can be minimised

✔ results of research into emergency procedures that you may have to take if something goes wrong.

You should also consider what to do at the end of the practical. For example, used agar plates should be left for a technician to sterilise; solutions of heavy metals should be collected in a bottle and disposed of safely.

Assessment tip

To make sure that your risk assessment is full and appropriate:

> Remember that, for a risk assessment for a chemical reaction, the risk assessment should be carried out for the products and the reactants.

> When using chemicals, make sure the hazard and ways of minimising risk match the concentration of the chemical you are using; many acids, for instance, while being corrosive in higher concentrations, are harmful or irritant at low concentrations.

✳ Collecting primary data

✔ You should make sure that any observations made are recorded in sufficient detail. For example, it is worth recording the appearance of a precipitate when making an insoluble salt, in addition to any other measurements that you make, such as its colour.

✔ Measurements should be recorded in tables. Have one ready so that you can record your readings as you carry out the practical work.

✔ Think about the dependent variable and define this carefully in your column headings.

✔ You should make sure that the table headings describe properly the type of measurements that you have made, for example 'time taken for magnesium ribbon to dissolve'.

✔ It is also essential that you include units – your results are meaningless without these.

✔ The units should appear in the column head, and not be repeated in each row of the table.

Definition

When you carry out an investigation, the data you collect are called **primary data.** The term 'data' is normally used to include your observations as well as measurements you might make.

Definition

One set of results from your investigation may not reflect what truly happens. Carrying out repeats enables you to identify any results that do not fit. These are called **outliers** or **anomalous results**.

Definition

If, when you carry out the same experiment several times, and get the same, or very similar results, the results are **repeatable**.

✳ Repeatability and reproducibility of results

When making measurements, in most instances, it is essential that you carry out repeats.

These repeats are one way of checking your results.

Results will not be repeatable of course, if you allow the conditions the investigation is carried out in to change.

You need to make sure that you carry out sufficient repeats, but not too many. In a titration, for example, if you obtain two values that are within $0.1\,cm^3$ of each other, carrying out any more will not improve the reliability of your results.

This is particularly important when scientists are carrying out scientific research and make new discoveries.

Definition

Taking more than one set of results will improve the **reliability** of your data.

Definition

The **reproducibility** of data is the ability of the results of an investigation to be reproduced by:

> using a different method and reaching the same conclusion

> someone else, who may be in a different lab, carrying out the same work.

✹ Processing data

Calculating the mean

Using your repeat measurements you can calculate the arithmetical mean (or just 'mean') of these data. Often, the mean is called the 'average.'

Here are the results of an investigation into the energy requirements of three different mp3 players. The students measured the energy using a joulemeter for ten seconds.

mp3 player	Energy used in joules (J)			
	trial 1	trial 2	trial 3	mean
Viking	5.5	5.3	5.7	5.5
Anglo	4.5	4.6	4.9	4.7
Saxon	3.2	4.5	4.7	4.6

Significant figures

When calculating the mean, you should be aware of significant figures.

For example, for the set of data below:

18	13	17	15	14	16	15	14	13	18

The total for the data set is 153, and ten measurements have been made. The mean is 15, and not 15.3.

This is because each of the recorded values has two significant figures. The answer must therefore have two significant figures. An answer cannot have more significant figures than the number being multiplied or divided.

Using your data

When calculating means (and displaying data), you should be careful to look out for any data that do not fit in with the general pattern. These are called anomalous data.

It might be the consequence of an error made in measurement, but sometimes it may be a genuine result. If you think that an anomalous result has been introduced by careless practical work, you should ignore it when calculating the mean. To check, the measurements may be repeated (If this is possible). You should examine possible reasons carefully before just ignoring data.

> **Definition**
>
> The **mean** is calculated by adding together all the measurements, and dividing by the number of measurements.

> **Assessment tip**
>
> You may also be required to use equations when processing data.
>
> Sometimes, you will need to rearrange an equation in order to make the calculation you need. Practise using and rearranging equations as part of your preparation for assessment.

> **Definition**
>
> **Significant figures** are the number of digits in a number based on the precision of your measurements.

Displaying your data

Displaying your data – usually the means – makes it easy to pick out and show any patterns. It also helps you to pick out any anomalous data.

It is likely that you will have recorded your results in tables, and you could also use additional tables to summarise your results. The most usual way of displaying data is to use graphs. The table will help you decide which type to use.

Type of graph	When you would use the graph	Example
bar chart or bar graph	where one of the variables is categoric	'the energy requirements of different mp3 players'
line graph	where independent and dependent variables are both continuous	'the volume of carbon dioxide produced by a range of different concentrations of hydrochloric acid'
scatter graph	to show an association between two (or more) variables	'the association between length and breadth of a number of privet leaves'

It should be possible, from the data, to either join the points of a line graph using a single straight line, or using a curve. In scatter graphs, the points are plotted, but not usually joined. In this way, graphs can also help you to understand the relationship between the independent variable and the dependent variable.

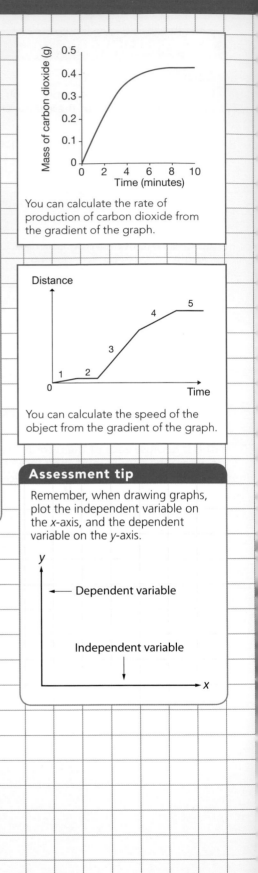

You can calculate the rate of production of carbon dioxide from the gradient of the graph.

You can calculate the speed of the object from the gradient of the graph.

Assessment tip

Remember, when drawing graphs, plot the independent variable on the x-axis, and the dependent variable on the y-axis.

✺ Conclusions from differences in data sets

When comparing two (or more) sets of data, you can often compare the values of two sets of means.

Example 5

Investigation: Two groups of students measured the amount of sweat that they produced after exercising for one hour. Their results are shown in the table below.

Student	Amount of sweat produced (ml)					Mean (ml)
	1	2	3	4	5	
Group 1	15	12	17	20	12	15.2
Group 2	11	25	12	13	10	11

When the means are compared, it appears that Group 1 produced more sweat than Group 2. The difference may have resulted from the amount and type of exercise that each group did, or it may be purely by chance.

Scientists use statistics to find the probability of any differences having occurred by chance. The lower this probability is, which is found out by statistical calculations, the more likely it is that tyre A is better at stopping a vehicle than tyre B.

Statistical analysis can help to increase the confidence you have in your conclusions.

Assessment tip

You have learned about probability in maths lessons.

Definition

If there is a relationship between dependent and independent variables that can be defined, there is a **correlation** between the variables.

✺ Drawing conclusions

Observing trends in data or graphs will help you to draw conclusions. You may obtain a linear relationship between two sets of variables, or the relationship might be more complex.

Example 6

Conclusion: As the temperature of the gas increased, its pressure also increased.

Conclusion: Increasing the temperature increased the energy of the gas particles, causing them to move around faster. This means that there were more collisions between the gas particles and the sides of the container and therefore the pressure increased.

When drawing conclusions, you should try to relate your findings to the science involved.

> In the first point in Example 6, your discussion should focus on describing what your results show, including any patterns or trends between them.

> In the second point in Example 6, there is a clear scientific mechanism to link the increase in temperature to an increase in gas pressure.

Sometimes, you can see correlations between data which are coincidental, where the independent variable is not the cause of the trend in the data.

This graph shows **negative correlation.**

This graph shows **positive correlation.**

✳ Evaluating your investigation

Your conclusion will be based on your findings, but must take into consideration any uncertainty in these introduced by any possible sources of error. You should discuss where these have come from in your evaluation.

The two types of errors are:

✔ random error
✔ systematic error.

This can occur when the instrument that you are using to measure lacks sufficient sensitivity to indicate differences in readings. It can also occur when it is difficult to make a measurement. If two investigators measure the height of a plant, for example, they might choose different points on the compost, and the tip of the growing point to make their measurements.

They are either consistently too high or too low. One reason could be down to the way you are making a reading, for example taking a burette reading at the wrong point on the meniscus. Another could be the result of an instrument being incorrectly calibrated, or not being calibrated.

The volume of liquid in a burette must be read to the bottom of the meniscus.

Definition
Error is a difference between a measurement you make and its true value.

Definition
With **random error**, measurements vary in an unpredictable way.

Definition
With **systematic error**, readings vary in a controlled way.

Assessment tip
A pH meter must be calibrated before use using buffers of known pH.

Assessment tip
Make sure that you relate your conclusions to the hypothesis you are investigating. Do the results confirm or reject the hypothesis. Quote some results to back up your statement. For example, 'My results at 35 °C and 65 °C show that over a 30 °C change in temperature the time taken to produce 50 cm^3 of carbon dioxide halved'

✳ Accuracy and precision

When evaluating your investigation, you should mention accuracy and precision. If you use these terms, it is important that you understand what they mean, and that you use them correctly.

Precise but not accurate.   Precise and accurate.   Not precise and not accurate.

The terms accuracy and precision can be illustrated using shots at a dartboard.

Definition
When making measurements:
> the **accuracy** of the measurement is how close it is to the true value

> **precision** is how closely a series of measurements agree with each other.

Improving your investigation

When evaluating your investigation, you should discuss how your investigation could be improved. This could be by improving:

✔ the reliability of your data. For example, you could make more repeats, or more frequent readings, or 'fine-tune' the range you chose to investigate, or refine your technique in some other way.

✔ the accuracy and precision of your data, by using measuring equipment with a higher resolution.

In science, the measurements that you make as part of your investigation should be as precise as you can, or need to, make them. To achieve this, you should use:

✔ the most appropriate measuring instrument

✔ the measuring instrument with the most appropriate size of divisions.

The smaller the divisions you work with, the more precise your measurements. For example:

✔ In an investigation on how your heart rate is affected by exercise, you might decide to investigate this after a 100 m run. You might measure out the 100 m distance using a trundle wheel, which is sufficiently precise for your investigation.

✔ In an investigation on how light intensity is affected by distance, you would make your measurements of distance using a metre rule with millimetre divisions; clearly a trundle wheel would be too imprecise.

✔ In an investigation on plant growth, in which you measure the thickness of a plant stem, you would use a micrometer or Vernier callipers. In this instance, a metre rule would be too imprecise.

Using secondary data

As part of controlled assessment, you will be expected to compare your data – primary data – with **secondary data** that others have collected.

One of the simplest ways of doing this is to compare your data with other groups in your class who have carried out an identical practical investigation.

In your controlled assessment, you will be provided with a data sheet of secondary data.

It may be that not all the investigations provided are relevant: different independent or dependent variables may have been investigated. It may be that some of the data has errors or uncertainties.

You should identify and explain, with reasons, whether each set of results can be used to support the hypothesis, and also any features of the data that might allow the hypothesis to be further developed, such as 'the effect is reduced with slopes of angle greater than 25°'.

Remember to use some figures from the results to support your answers.

Definition

Secondary data are measurements/observations made by anyone other than you.

Assessment tip

In Section 2 of your ISA you will be required to critically evaluate this data to see how far it supports or does not support the hypothesis being investigated.

You should review secondary data and evaluate it. Scientific studies are sometimes influenced by the **bias** of the experimenter.

✔ One kind of bias is having a strong opinion related to the investigation, and perhaps selecting only the results that fit with a hypothesis or prediction.

✔ The bias could be unintentional. In fields of science that are not yet fully understood, experimenters may try to fit their findings to current knowledge and thinking.

There have been other instances where the 'findings' of experimenters have been influenced by organisations that supplied the funding for the research.

You must fully reference any secondary data that you have used, using one of the accepted referencing methods.

✳ Referencing methods

The two main conventions for writing a reference are the:

✔ Harvard system
✔ Vancouver system.

In your text, the Harvard system refers to the authors of the reference, for example 'Smith and Jones (1978)'.

The Vancouver system refers to the number of the numbered reference in your text, for example '... the reason for this hypothesis is unknown[1]'.

Though the Harvard system is usually preferred by scientists, it is more straightforward for you to use the Vancouver system.

Harvard system

In your references list a book reference should be written:

> Author(s) (year of publication). *Title of Book*, publisher, publisher location.

The references are listed in alphabetical order according to the authors.

Vancouver system

In your references list a book reference should be written:

> 1 Author(s). *Title of Book*. Publisher, publisher location: year of publication.

The references are numbered in the order in which they are cited in the text.

Assessment tip

Remember to write out the URL of a website in full. You should also quote the date when you looked at the website.

✳ Do the data support your hypothesis?

You need to discuss, in detail, whether all, or some, of the primary data that you have collected, and the secondary data, support your original hypothesis. They may or may not.

You should communicate your points clearly, using the appropriate scientific terms.

If your data do not match your hypothesis completely, it may be possible to modify the hypothesis or suggest an alternative one.

You may be asked to suggest further investigations that can be carried out to support your original hypothesis or the modified version.

It is important to remember, however, that if your investigation does support your hypothesis, it can improve the confidence you have in your conclusions and scientific explanations, but it cannot prove that your explanations are correct. One result can never prove a theory, but it can disprove it.

✳ Your controlled assessment

The **Controlled Assessment** is worth 25% of the marks for each Science subject. It is worth doing it well!

Controlled Assessment is a two-part (sections 1 and 2) *Investigative Skills Assignment* (ISA) test. Before you start, you will discuss a problem in a given context. For example, how does a bungee jumping company know how much rope to use to avoid jumpers hitting the ground?

> You will then research variables that affect the stretch of a bungee rope, and devise a method to test the stretch of a bungee rope.

> You will also suggest a hypothesis about the stretch of a bungee rope as the result of your research.

You are allowed to make brief notes about your research on one side of A4. You can use the notes to help answer Sections 1 and 2 of the ISA paper.

Section 1 (45 minutes, 20 marks) consists of questions relating to your research. This includes your planned investigation. Then you will carry out your investigation and record and analyse your results.

Section 2 of the ISA test (50 minutes, 30 marks) consists of questions related to your investigation. You will be given a sheet of secondary data by AQA. You should use it to select data to analyse and compare with the hypothesis. Finally, you will be asked to suggest how ideas from your investigation and research could be used in a specific context.

Assessment tip

Write your plan clearly, using the appropriate scientific terms, and checking carefully your use of spelling, punctuation and grammar. You will be assessed on this written communication as well as your science.

How to be successful in GCSE Separate Sciences assessment

Introduction

AQA uses assessments to test how good your understanding of scientific ideas is, how well you can apply your understanding to new situations and how well you can analyse and interpret information you have been given. The assessments are opportunities to show how well you can do these.

To be successful in exams you need to:

✔ have a good knowledge and understanding of science
✔ be able to apply this knowledge and understanding to familiar and new
✔ situations
✔ be able to interpret and evaluate evidence that you have just been given.

You need to be able to do these things under exam conditions.

✹ The language of the assessment paper

When working through an assessment paper, make sure that you:

✔ re-read a question enough times until you understand exactly what the examiner is looking for
✔ make sure that you highlight key words in a question. In some instances, you will be given key words to include in your answer.
✔ look at how many marks are allocated for each part of a question. In general, you need to write at least as many separate points in your answer as there are marks.

✹ What verbs are used in the question?

A good technique is to see which verbs are used in the wording of the question and to use these to gauge the type of response you need to give. The table lists some of the common verbs found in questions, the types of responses expected and then gives an example.

Verb used in question	Response expected in answer	Example question
write down state give identify	these are usually more straightforward types of question in which you are asked to give a definition, make a list of examples, or the best answer from a series of options	'Write down three types of microorganism that cause disease.' 'State one difference and one similarity between radio waves and gamma rays.'
calculate	use maths to solve a numerical problem	'Calculate the percentage of carbon in copper carbonate $(CuCO_3)$.'

estimate	use maths to solve a numerical problem, but you do not have to work out the exact answer	'Estimate from the graph the speed of the vehicle after 3 minutes.'
describe	use words (or diagrams) to show the characteristics, properties or features of, or build an image of something	'Describe how meiosis halves the number of chromosomes in a cell to make egg or sperm cells.'
suggest	usually in a new or unfamiliar context	'Suggest why tyres with different tread patterns will have different braking distances.'
demonstrate/ show how	use words to make something evident using reasoning	'Show how temperature can affect the rate of a chemical reaction.'
compare	look for similarities and differences	'Compare aerobic and anaerobic respiration.'
explain	to offer a reason for, or make understandable, information that you are given	'Explain why alpha and beta radiations can be deflected by a magnetic field, but gamma rays are not.'
evaluate	discuss different points of view or opinions, then decide which is best, usually giving a reason	'Evaluate the benefits of using a circuit breaker instead of a fuse in an electrical circuit.'

✳ What is the style of the question?

Try to get used to answering questions that have been written in lots of different styles before you sit the exam. Work through past papers, or specimen papers, to get a feel for these. The types of questions in your assessment fit the three assessment objectives shown in the table.

Assessment objective	Your answer should show that you can…
AO1 recall the science	Recall, select and communicate your knowledge and understanding of science.
AO2 apply your knowledge	Apply skills, knowledge and understanding of science in practical and other contexts.
AO3 evaluate and analyse the evidence	Analyse and evaluate evidence, make reasoned judgements and draw conclusions based on evidence.

Assessment tip

Of course you must revise the subject material adequately. It is as important that you are familiar with the different question styles used in the exam paper, as well as the question content.

✹ How to answer questions on: AO1 Recall the science

These questions, or parts of questions, test your ability to recall your knowledge of a topic. There are several types of this style of question:

✔ Fill in the spaces (you may be given words to choose from)
✔ Tick the correct statements
✔ Use lines to link a term with its definition or correct statement
✔ Add labels to a diagram
✔ Complete a table
✔ Describe a process

Example 7

a Which is the correct equation to calculate the pressure that an object exerts?

Tick (✓) **one** box.

☐ $P = \dfrac{F}{A}$

☐ $P = \dfrac{A}{F}$

☐ $P = F \times A$

✹ How to answer questions on: AO1 Recall the science in practical techniques

You may be asked to recall how to carry out certain practical techniques, either ones that you have carried out before, or techniques that scientists use.

To revise for these types of questions, make sure that you have learned definitions and scientific terms. Produce a glossary of these, or key facts cards, to make them easier to remember. Make sure that your key facts cards also cover important practical techniques, including equipment, where appropriate.

Assessment tip

Don't forget that mind maps – either drawn by you or by using a computer program – are very helpful when revising key points.

Example 8

Describe two factors that scientists can change to affect the amount of product produced in an equilibrium reactions such as the Haber process or Contact process.

✹ How to answer questions on: AO2 Apply skills, knowledge and understanding

Some questions require you to apply basic knowledge and understanding in your answers.

You may be presented with a topic that is familiar to you, but you should also expect questions in your Science exam to be set in an unfamiliar context.

Questions may be presented as:

✔ practical investigations
✔ data for you to interpret
✔ a short paragraph or article.

The information required for you to answer the question might be in the question itself, but for later stages of the question, you may be asked to draw on your knowledge and understanding of the subject material in the question.

Practice will help you to become familiar with contexts that examiners use and question styles. However, you will not be able to predict many of the contexts used. This is deliberate; being able to apply your knowledge and understanding to different and unfamiliar situations is a skill the examiner tests.

Practise doing questions where you are tested on being able to apply your scientific knowledge and your ability to understand new situations that may not be familiar. In this way, when this type of question comes up in your exam, you will be able to tackle it successfully.

Example 9

When light enters the eye from the air, it passes first through the cornea – a jelly-like substance – before entering the pupil. Use your knowledge of refraction to explain what happens to the light as it passes through the cornea.

Assessment tip

Work through the Preparing for assessment: Applying your knowledge tasks in this book as practice.

✳ How to answer questions on: AO2 Apply skills, knowledge and understanding in practical investigations

Some opportunities to demonstrate your application of skills, knowledge and understanding will be based on practical investigations. You may have carried out some of these investigations, but others will be new to you and based on data obtained by scientists. You will be expected to describe patterns in data from graphs you are given or that you will have to draw from given data.

Again, you will have to apply your scientific knowledge and understanding to answer the question.

Example 10

A student measured the pH of two different brands of beer over seven days, to compare how quickly the ethanol in each beer was oxidised to ethanoic acid. His results are shown in the table on the right.

a   Calculate the total pH decrease for each beer over the six days.

b   Plot the results as a line graph.

c   Ethanoic acid has a pH of 2.4. Use the graph to work out which beer oxidised to ethanoic acid first.

Beer	pH on each day					
	1	2	3	4	5	6
Old Brew	6.6	5.8	4.0	3.0	2.1	2.1
Eagle larger	6.3	5.9	4.1	3.5	2.2	2.1

You will also need to analyse scientific evidence or data given to you in the question. It is likely that you will not be familiar with the material.

Analysing data may involve drawing graphs and interpreting them, and carrying out calculations. Practise drawing and interpreting graphs from data.

When drawing a graph, make sure that you:

✔ choose and label the axes fully and correctly

✔ include units, if this has not been done already

✔ plot points on the graph carefully – the examiner will check individual points to make sure that they are accurate

✔ join the points correctly; usually this will be by a line of best fit.

When reading values off a graph that you have drawn or one given in the question, make sure that you:

✔ do it carefully, reading the values as accurately as you can

✔ double-check the values.

When describing patterns and trends in the data, make sure that you:

✔ write about a pattern or trend in as much detail as you can

✔ mention anomalies where appropriate

✔ recognise there may be one general trend in the graph, where the variables show positive or negative correlation

✔ recognise the data may show a more complex relationship. The graph may demonstrate different trends in several sections. You should describe what is happening in each.

✔ describe fully what the data show.

✹ How to answer questions needing calculations

✔ The calculations that you are asked to do may be straightforward, for example the calculation of the mean from a set of data.

✔ They may be more complex, for example calculating the yield of a chemical reaction.

✔ Other questions will require the use of formulae.

You will be given an equation sheet with the question paper.

On page 299, there is a list of the maths skills that you will need. Remember, these are the same skills that you have learned in maths lessons.

Example 11

10.0 cm³ of a solution of sodium hydroxide was titrated with a 0.10 M solution of hydrochloric acid. Calculate the concentration of the sodium hydroxide solution if 15.0 cm³ of hydrochloric acid was needed to neutralise it.

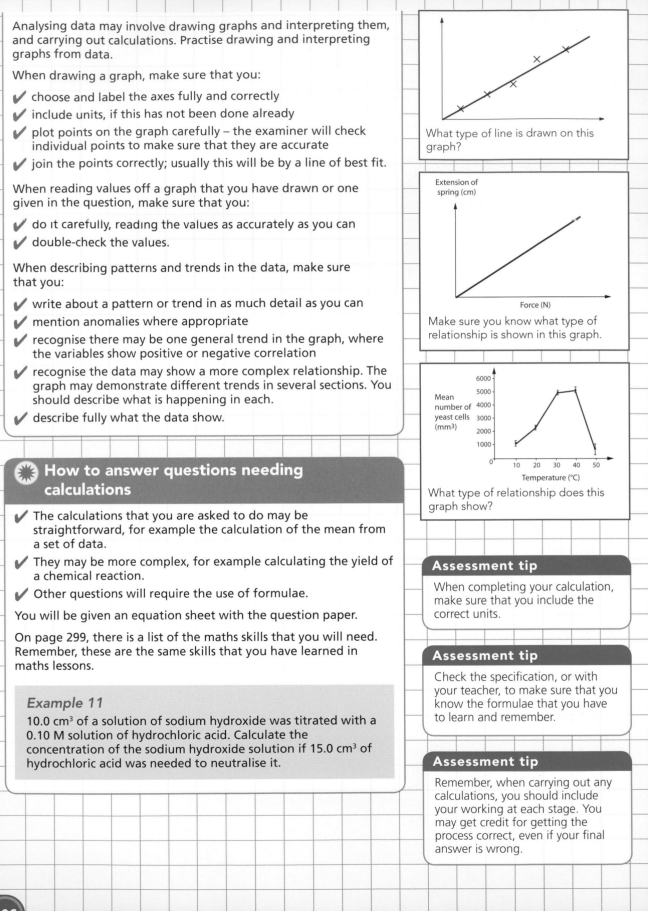

What type of line is drawn on this graph?

Extension of spring (cm) / Force (N)

Make sure you know what type of relationship is shown in this graph.

What type of relationship does this graph show?

Assessment tip

When completing your calculation, make sure that you include the correct units.

Assessment tip

Check the specification, or with your teacher, to make sure that you know the formulae that you have to learn and remember.

Assessment tip

Remember, when carrying out any calculations, you should include your working at each stage. You may get credit for getting the process correct, even if your final answer is wrong.

✳ How to answer questions on: AO3 Analysing and evaluating evidence

For these types of questions, in addition to analysing data, you must also be able to evaluate information that you are given. This is one of the hardest skills. Think about the validity of the scientific data: did the technique(s) used in any practical investigation allow the collection of accurate and precise data?

Your critical evaluation of scientific data in class, along with the practical work and controlled assessment work, will help you to develop the evaluation skills required for these types of questions.

Example 12

Why might there be a difference in the data collected in an investigation to measure the mass lost in the reaction between marble chips and acid, if some reactions were carried out in conical flasks and others in beakers?

You may be expected to compare data with other data, or come to a conclusion about its reliability, its usefulness or its implications. Again, it is possible that you will not be familiar with the context. You may be asked to make a judgement about the evidence or to give an opinion with reasons.

Example 13

When investigating the activity of an enzyme, why is it hard to achieve repeatable results when the enzyme is used at temperatures above 40 °C?

> **Assessment tip**
>
> Work through the Preparing for assessment: Analysing and evaluating data tasks, in this book, as practice.

> **Assessment tip**
>
> Wherever possible, use as much data as you can in your answer, particularly when explaining trends or conclusions, so you can gain full marks. Try to use numbers and values rather than just trends in data or graphs. 'At 45 °C…' is always better than 'as the temperature rises it gets greater'.

✳ The quality of your written communication

Scientists need good communication skills to present and discuss their findings. You will be expected to demonstrate these skills in the exam. You will be assessed in the longer-response exam questions that you answer. These questions are clearly indicated in each question paper. The quality of your written communication will also be assessed in your controlled assessment.

You will not be able to obtain full marks unless you:

✔ make sure that the text you write is legible
✔ make sure that spelling, punctuation and grammar are accurate so that the meaning of what you write is clear
✔ use a form and style of writing appropriate for its purpose and for the complexity of the subject matter
✔ organise information clearly and coherently
✔ use the scientific language correctly.

You will also need to remember the writing and communication skills that you have developed in English lessons. For example, make sure that you understand how to construct a good sentence using connectives.

> **Assessment tip**
>
> You will be assessed on the way in which you communicate science ideas.

> **Assessment tip**
>
> When answering questions, you must make sure that your writing is legible. An examiner cannot award marks for answers that he or she cannot read.

✳ Revising for your exam

You should revise in the way that suits you best. It is important that you plan your revision carefully, and it is best to start well before the date of the exams. Take the time to prepare a revision timetable and try to stick to it. Use this during the lead up to the exams and between each exam.

When revising:

✔ Find a quiet and comfortable space in the house where you will not be disturbed. It is best if it is well ventilated and has plenty of light.

✔ Take regular breaks. Some evidence suggests that revision is most effective when you revise in 30 to 40 minute slots. If you get bogged down at any point, take a break and go back to it later when you're feeling fresh. Try not to revise when you are feeling tired. If you do feel tired, take a break.

✔ Use your school notes, textbook and possibly a revision guide. But also make sure that you spend some time using past papers to familiarise yourself with the exam format.

✔ Produce summaries of each topic.

✔ Draw mind maps covering the key information on a topic.

✔ Set up revision cards containing condensed versions of your notes.

✔ Ask yourself questions, and try to predict questions, as you are revising a topic.

✔ Test yourself as you go along. Try to draw key labelled diagrams, and try some questions under timed conditions.

✔ Prioritise your revision of topics. You might want to allocate more time to revising the topics you find most difficult.

Assessment tip

Try to make your revision timetable as specific as possible – don't just say 'science on Monday, and Thursday', but list the topics that you will cover on those days.

Assessment tip

Start your revision well before the date of the exams, produce a revision timetable, and use the revision strategies that suit your style of learning. Above all, revision should be an active process.

✳ How do I use my time effectively in the exam?

Timing is important when you sit an exam. Do not spend so long on some questions that you leave insufficient time to answer others. For example, in a 60-mark question paper, lasting one hour, you will have, on average, one minute per question.

If you are unsure about certain questions, complete the ones you are able to do first, then go back to the ones you're less sure of.

If you have time, go back and check your answers at the end of the exam.

✳ On exam day...

A little bit of nervousness before your exam can be a good thing, but try not to let it affect your performance in the exam. When you turn over the exam paper keep calm. Look at the paper and get it clear in your head exactly what is required from each question. Read each question carefully. Do not rush.

If you read a question and think that you have not covered the topic, keep calm – it could be that the information needed to answer the question is in the question itself or the examiner may be asking you to apply your knowledge to a new situation.

Finally, good luck!

✳ Mathematical skills

You will be allowed to use a calculator in all assessments.

These are the maths skills that you need, to complete all the assessments successfully.

You should understand:

✔ the relationship between units, for example, between a gram, kilogram and tonne
✔ compound measures such as speed
✔ when and how to use estimation
✔ the symbols = < > ~
✔ direct proportion and simple ratios
✔ the idea of probability.

You should be able to:

✔ give answers to an appropriate number of significant figures
✔ substitute values into formulae and equations using appropriate units
✔ select suitable scales for the axes of graphs
✔ plot and draw line graphs, bar charts, pie charts, scatter graphs and histograms
✔ extract and interpret information from charts, graphs and tables.

You should be able to calculate:

✔ using decimals, fractions, percentages and number powers, such as 10^3
✔ arithmetic means
✔ areas, perimeters and volumes of simple shapes

In addition, if you are a higher tier candidate, you should be able to:

✔ **change the subject of an equation**

and should be able to use:

✔ **numbers written in standard form**
✔ **calculations involving negative powers, such as 10^{-1}**
✔ **inverse proportion**
✔ **percentiles and deciles.**

✳ Some key physics equations

With the written papers, there will be an equation sheet. In order to make best use of the sheet, it will help if you practise using the following equations.

Equation	Meaning of symbol and its unit
$s = v \times t$	s is distance in metres, m v is speed in metres per second, m/s t is time in seconds, s
$n = \dfrac{\sin i}{\sin r}$	n is refractive index i is angle of incidence r is angle of refraction
$n = \dfrac{1}{\sin c}$	n is refractive index c is critical angle
$P = \dfrac{1}{f}$	P is power in dioptres, D f is focal length in metres, m
$T = \dfrac{1}{f}$	T is periodic time in seconds, s f is frequency in hertz, Hz
$M = F \times d$	M is moment of the force in newton metres, N m F is force in newtons, N d is the perpendicular distance from the line of the action of the force to the pivot in metres, m
$P = \dfrac{F}{A}$	P is pressure in pascals, Pa F is force in newtons, N A is cross sectional area in metres squared, m²
$W = F \times d$	W is work done in joules, J F is force applied in newtons, N d is distance moved in the direction of the force in metres, m
$\dfrac{V_p}{V_s} = \dfrac{n_p}{n_s}$	V_p is potential difference across the primary coil in volts, V V_s is potential difference across the secondary coil in volts, V n_p is the number of turns on the primary coil n_s is the number of turns on the secondary coil
$V_p \times I_p = V_s \times I_s$	V_p is potential difference across the primary coil in volts, V I_p is current in the primary coil in amperes (amps), A V_s is potential difference across the secondary coil in volts, V I_s is current in the secondary coil in amperes (amps), A

Glossary

acceleration   rate at which an object speeds up, calculated from velocity divided by time

activation energy   minimum energy needed to break bonds in reactant molecules to allow a reaction to occur

active transport   movement, through a cell membrane, of a substance against its concentration gradient, using energy supplied by the cell

alveoli   tiny air sacs in lungs, where gas exchange takes place between air and blood

antigens   proteins, on the surface of cells, recognised as foreign by the immune system

aorta   largest artery in the body – it carries blood from the left ventricle to the body

arteries   thick-walled blood vessels that carry high-pressure blood away from the heart

arterioles   small arteries

atomic number   number of protons in the nucleus of an atom

atria   upper chambers of the heart – they receive blood from the veins

Avogadro's number   number of particles contained in one mole – 6.02×10^{23}

biodiesel   fuel made from plant oils such as rapeseed

biodiversity   range of different living organisms in a habitat

bioethanol   ethanol (alcohol) made by the action of microorganisms on sugars

biogas   mixture of methane and carbon dioxide made by the action of microorganisms on organic material

blood vessels   tubes through which blood travels

bond energy   amount of energy absorbed to break one mole of bonds or given out when one mole of bonds form

bromide   salt formed when a metal reacts with bromine – bromide ions have a -1 charge

burette   item of glassware to measure an exact volume of reactant during a titration

calorie   one calorie is the amount of energy needed to raise the temperature of 1 g water by 1 °C

calorimetry   method to measure the energy given out in combustion reactions

capillaries   the smallest blood vessels – they distribute blood close to every cell

carbon   atomic number 6 – used in water filters to absorb impurities

catalyst   substance added to a chemical reaction to alter (usually increase) the reaction rate, without being used up in the process

centre of mass   position where all the mass would end up if an object collapsed inwards to a point

centripetal force   force acting towards the centre of a circle that keeps an object moving in a circle

chloride   salt formed when chlorine reacts with a metal – chloride ions have a -1 charge

ciliary muscles   muscles that pull on the lens in an eye – they focus the lens by changing its shape

closed system   environment for a chemical reaction where no reactants or products can escape

compress   squashing something into a smaller volume – liquids are incompressible: they cannot be squashed

concave lens   lens that curves inwards with the centre thinner than the edges – it causes light rays to diverge

concentration gradient   a difference in the concentration of ions across a cell membrane

constrict   become narrower

converging lens   convex lens that causes light rays to come together

convex lens   lens that curves outwards with the centre thicker than the edges – it causes light rays to converge

core body temperature   temperature deep inside the body

cornea   the clear surface that allows light into the eye and provides most of the focusing

critical angle   angle of incidence that gives a refracted angle of 90° – at angles of incidence greater than the critical angle, light will be totally internally reflected

CT (or CAT) scan   scan that takes many X-ray images from different directions to build up a 3-D image of the body

deoxygenated (blood)   containing only a little oxygen

desalination   method to remove dissolved salts from seawater

dialysis   separating substances with different sizes of molecules or ions, using a partially permeable membrane

digital camera   camera where the light is recorded on a semiconductor material rather than on photographic film

dilate   become wider

dioptre   unit to measure the power of a lens

displacement reaction   chemical reaction in which one atom or group of atoms in a compound is replaced by another

distillation   process for separating liquids by boiling them, then condensing the vapours

diverging lens   concave lens that causes light rays to spread out

donor   person who provides an organ for transplant

Doppler changes   changes in wavelength and frequency of a wave reflected from a moving object

dynamic equilibrium   when the rate of the forward reaction is equal to the rate of the backward reaction in a reversible reaction

echo   reflection of a sound or ultrasound wave

efficiency   useful energy output divided by the total energy input – often expressed as a percentage

electrical power   amount of energy transferred each second and measured in watts or joules per second

electromagnet magnet formed by an electric current flowing through a solenoid – the magnetic field of an electromagnet can be switched on and off, by switching the current on and off

electromagnetic (EM) spectrum electromagnetic waves ordered according to wavelength and frequency – ranging from radio waves to gamma rays

electromagnetic induction potential difference across and current through a coil, caused by a changing magnetic field

end point when exact amounts of reactants have been added to obtain a neutral solution in a titration

endoscope device containing an optical fibre, a light and a camera, used to examine inside patients

endothermic chemical reaction which takes in heat or energy from other sources

energy level diagram diagram that shows how the energy content of reactants and products changes during a chemical reaction

exchange surface part of the body where substances move between the body and its environment

exothermic reaction in which energy is given out – energy is transferred from the reactants to the surroundings

far point the furthest point at which the eye can focus clearly – normally this is at infinity

flame test experimental technique used to identify mainly Group 1 and 2 metal ions by observing the colour of light emitted when samples are heated

Fleming's left-hand rule shows which way a current-carrying wire tries to move when placed in a magnetic field – First finger shows Field, seCond finger shows Current, thuMb shows Movement.

focal length distance between the centre point of a lens and the point (the principal focus) where parallel rays entering the lens either cross over, or appear to have come from

force multiplier device that reduces the size of the force needed to move something

frequency number of waves passing a set point in one second

frequency (pendulum) number of times for a pendulum to pass from its highest point on one side and back to the same point in one second

friction force acting at points of contact between objects moving over each other, to resist the movement

functional group an atom or group of atoms that characterise a homologous series of organic compounds

global warming increase in the mean temperature on Earth

glucagon hormone secreted by the pancreas when blood glucose levels fall too low – it brings about actions to increase blood glucose level

glycogen substance made from glucose molecules linked in a long chain – found in the liver and muscles, where it forms a glucose store

gravitational potential energy energy that an object has because of its position, for example, increasing the height of an object above the ground increases its gravitational potential energy

gravity, force due to the attractive force acting between all objects with mass – it pulls objects downwards on Earth and keeps planets and satellites in orbit

group a vertical column in the periodic table

Haber process industrial process to manufacture ammonia from hydrogen gas and nitrogen gas

haemoglobin red protein found inside red blood cells – it combines reversibly with oxygen

halide salt formed when a halogen reacts with a metal

homeostasis the maintenance of a constant internal environment in the body

homologous series family of organic compounds with the same functional group

hydraulic system system of liquid in pipes that can transmit force from one place to another and can also act as a force multiplier

hydroxyl the OH group

hyphae tiny, thread-like structures that make up the body of a fungus

immunosuppressant drugs medicine that reduces the activity of the immune system

indicator substance used to detect pH in a titration

insulin hormone secreted by the pancreas when blood glucose levels rise too high – it brings about actions to reduce blood glucose level

internal environment conditions in which cells inside the body operate

inverted an inverted image is upside down compared with the object

iodide salt formed when iodine reacts with a metal

ion exchange column apparatus used to purify and/or soften water

ion exchange resin material used in ion exchange columns – it exchanges unwanted ions in water for hydrogen or sodium ions

ionisation causing electrons to split away from their atoms – some radiation is harmful to living cells because it is ionising

isotonic drink a drink that has the same concentration of different ions, glucose and water as blood

joule unit used to measure energy

kinetic energy energy an object has because of its movement – it is greater for objects with greater mass or higher speed

laser device producing a very narrow, highly focused, high energy beam of single wavelength electromagnetic radiation

Le Chatelier's principle if a dynamic equilibrium is disturbed by changing the conditions, the position of equilibrium moves to counteract the change.

lens piece of curved glass or plastic designed to refract light in a specific way

lever   device that uses the moment of a force to move something – many levers are also force multipliers

long sight   vision defect where people cannot see near objects clearly – corrected by using a convex lens

lungs   organs in the thorax within which gas exchange occurs

magnetic field   area around a magnet or current-carrying wire, where there is a force on magnetic objects or current carrying wires.

magnification   measure of how much larger an image is than the object – if the image is smaller than the object, the magnification is less than 1

medium (pl. media)   the material through which light or other types of wave travel

moles   unit for counting atoms and molecules – one mole of any substance contains the same number of particles

moment   turning effect of a force – moment is increased by increasing the force or the distance between force and pivot

motor effect   interaction between a magnetic field due to a magnet and a current-carrying wire that causes movement of the wire

mycoprotein   a protein-rich food produced by the fungus *Fusarium*

National Grid   network that distributes electricity from power stations across the country

near point   the closest point at which the eye can focus clearly – for a young adult this is normally about 25 cm

normal   line at right angles to a boundary – used to help draw ray diagrams

optical fibre   glass fibre that is used to transfer signals as light or infrared radiation

oscilloscope   device with screen to show how amplitude and frequency of an input wave varies – also called a cathode ray oscilloscope

osmosis   the diffusion of water from a dilute solution (where there is a lot of water) to a concentrated solution (where there is less water), through a partially permeable membrane

oxygenated (of blood)   containing a lot of oxygen

oxyhaemoglobin   haemoglobin that has combined with oxygen

pancreas   organ lying to the left of the stomach – it secretes pancreatic juice (containing digestive enzymes) and the hormones insulin and glucagon

partially permeable membrane   membrane with extremely tiny holes that lets small molecules, such as water, pass through, but not larger molecules, such as sugar

pendulum   object, suspended from a point, that swings to and fro moving along part of a circular path

period   horizontal row in the periodic table

periodic time   time taken for one complete swing (back and forth) of a pendulum

permanent hard water   water containing dissolved calcium sulfate and/or dissolved magnesium sulfate

phloem   tissue made up of long tubes that transport sugars from the leaves or storage organs to all other parts of a plant

pipette   apparatus used to measure a set volume of liquid

pivot   point around which a lever or a seesaw turns

planet   large ball of gas or rock travelling around a star – for example Earth and other planets orbit our Sun

plaques   deposits of cholesterol that form in the walls of arteries

plasma   liquid part of blood

platelet   a tiny structure found in blood and that helps with clotting

position of equilibrium   the relative amounts of reactants and products present when a state of dynamic equilibrium is reached in a reversible reaction

potential difference   difference in electrical potential between two points in a circuit – also called the voltage between two points

power (of a lens)   measure of how much a lens makes light change direction

power transmission cables   wires used by the National Grid to transmit electrical energy from a power station to homes and businesses

precipitate   solid product formed by reacting two solutions

pressure   pressure at any point is the force acting at that point divided by the area over which the force acts – pressure increases if force increases or area decreases

primary coil   coil of a transformer across which the input potential difference is connected

principal axis   line passing through the centre of a lens, and perpendicular to the plane of symmetry of the lens – the principal focus of the lens is always on the principal axis

principal focus   point at which parallel rays entering a lens either cross over, or appear to have come from

pupil   place where light enters the front of the eye – the pupil changes size to control the amount of light entering the eye

range of vision   range of distances over which the eye can focus light clearly – for normal vision it is from about 25 cm to infinity

ray diagram   line diagram showing how light rays travel

reabsorption   taking back useful substances into the blood from the liquid that has been filtered in the kidneys

recipient   person who receives a transplanted organ

red blood cells   the most numerous cells in blood – they transport oxygen

refraction   change of direction of a wave when it hits a boundary between two different media at an angle, for example when a light ray passes from air into a glass block

refractive index   measure of how much a medium refracts light

rejection (of transplant)   an attack on a transplanted organ by the recipient's immune system

relative formula mass   combined mass of all atoms in a formula – for each type of atom, multiply its relative atomic mass by the number of atoms present

resultant force   the single force that would have the same effect on an object as all the forces that are acting on the object

resultant moment   the single moment that would have the same effect as all the moments acting on an object

retina   light-sensitive surface at the back of the eye that sends electrical signals to the brain

ribcage   protective structure formed by the ribs, enclosing the heart and lungs within the thorax

satellite   any natural or artificial object orbiting around a larger object

secondary coil   coil of a transformer across which an output potential difference is induced

sequestration   locking away – for example, peat bogs sequester carbon

short sight   vision defect where people cannot see distant objects clearly – corrected by using a concave lens

silver   atomic number 47, used in water filters to kill microbes

soft iron core   laminated core of soft iron around which the coils of a transformer are wound – the current in the primary coil causes a magnetic field in the soft iron core

solenoid   a coil of current-carrying wire that generates a magnetic field

specific heat capacity   the energy needed to raise the temperature of one gram of a substance by one degree Celsius

speed   rate at which an object is moving

speed of light   speed at which electromagnetic radiation travels through a vacuum – 300 000 000 metres per second

stability   stability indicates how easily an object will topple over – the wider the base and the lower the centre of mass, the more stable an object is, and the less likely to topple

steam reforming   chemical process in which methane is reacted with steam to make carbon dioxide and hydrogen

stent   a tiny tube inserted into an artery to keep it open

step-down transformer   transformer that changes alternating current to a lower voltage

step-up transformer   transformer that changes alternating current to a higher voltage

stoma (pl. stomata)   tiny hole in the surface of a leaf, through which gases can diffuse

suspensory ligaments   ligaments that hold the lens in place in the eye

sustainable food production   producing or harvesting food in a way that ensures there will be plenty left in the future

sweat glands   organs in the skin that secrete sweat to help reduce body temperature

switch mode transformer   small, light, modern transformer that operates using a very high frequency alternating current

Table of Octaves   an early periodic table devised by John Newlands

temperature receptors   sense organs that detect changes in temperature

temporary hard water   water containing dissolved calcium hydrogen carbonate

tension   force that pulls or stretches

thermoregulatory centre   the part of the brain that controls temperature regulation

thorax   the part of the body between the head and the abdomen – it is separated from the abdomen by the diaphragm

time period (of pendulum)   time taken for one complete swing (back and forth) of a pendulum. Also called the periodic time of the pendulum.

titration   procedure to determine the volume of one solution needed to react with a known volume of another solution

total internal reflection   when a ray of light that hits the boundary between glass and air is reflected back into the glass, so that none of it leaves the glass block or glass fibre

transpiration   loss of water vapour from leaves

transpiration stream   movement of water through a plant, upwards from its roots and eventually out through the stomata in the leaves

transplant   transfer of a body organ from one person to another

Type 1 diabetes   a condition in which the pancreas does not secrete sufficient insulin

ultrasound   sound waves that have a frequency too high for humans to hear

urea   excretory product produced in the liver from excess amino acids

urine   solution containing urea, produced by the kidneys

veins   thin-walled blood vessels that carry low-pressure blood towards the heart

velocity   measure of how fast an object is moving in a particular direction

ventilation   breathing movements, which move air into and out of the lungs

ventricles   the lower chambers of the heart that pump blood out to the body

villus (pl. villi)   tiny, finger-like projection on the inner surface of the small intestine – villi increase the surface area for absorption of nutrients

virtual image   image that can be seen but cannot be projected onto a screen (mirrors and concave lenses form virtual images)

virtual ray   line showing where a ray of light appears to have come from – drawn as dotted lines on ray diagrams

washing soda   common name for sodium carbonate

wavelength   distance between two points of maximum amplitude on a wave

white blood cells   cells, found in blood, that help to fight against disease-causing bacteria and viruses

X-rays   ionising electromagnetic radiation – used in X-ray photography to generate picture of bones or teeth, and in CT scans.

xylem   tissue made up of long tubes that transport water and mineral ions from the roots to all parts of a plant

Index

✳ Internet research

The internet is a great resource to use when you are working through your GCSE Science course.

Below are some tips to make the most of it.

1   Make sure that you get information at the right level for you by typing in the following words and phrases after your search: 'GCSE', 'KS4', 'KS3', 'for kids', 'easy', or 'simple'.

2   Use OR, AND, NOT and NEAR to narrow down your search.

> Use the word OR between two words to search for one or the other word.

> Use the word AND between two words to search for both words.

> Use the word NOT, for example, 'York NOT New York' to make sure that you do not get unwanted results (hits).

> Use the word NEAR, for example, 'London NEAR Art' to bring up pages where the two words appear very close to each other.

3   Be careful when you search for phrases. If you search for a whole phrase, for example, A Room with a View, you may get a lot of search results matching some or all of the words. If you put the phrase in quote marks, 'A Room with a View' it will only bring search results that have that whole phrase and so bring you more pages about the book or film and less about flats to rent!

4   For keyword searches, use several words and try to be specific. A search for 'asthma' will bring up thousands of results. But, a search for 'causes of asthma' or 'treatment of asthma' will bring more specific and fewer returns. Similarly, if you are looking for information on cats, for example, be as specific as you can by using the breed name.

5   Most search engines list their hits in a ranked order so that results that contain all your listed words (and so most closely match your request) will appear first. This means the first few pages of results will always be the most relevant.

6   Avoid using lots of smaller words such as A or THE unless it is particularly relevant to your search. Choose your words carefully and leave out any unnecessary extras.

7   If your request is country-specific, you can narrow your search by adding the country. For example, if you want to visit some historic houses and you live in the UK, search 'historic houses UK' otherwise it will search the world. With some search engines you can click on a 'web' or 'pages from the UK only' option.

8   Use a plus sign (+) before a word to force it into the search. That way only hits with that word will come up.

Modern periodic table

Group 1	2												3	4	5	6	7	0
1 1 H hydrogen																		4 2 He helium
7 3 Li lithium	9 4 Be beryllium												11 5 B boron	12 6 C carbon	14 7 N nitrogen	16 8 O oxygen	19 9 F fluorine	20 10 Ne neon
23 11 Na sodium	24 12 Mg magnesium												27 13 Al aluminium	28 14 Si silicon	31 15 P phosphorus	32 16 S sulfur	35 17 Cl chlorine	40 18 Ar argon
39 19 K potassium	40 20 Ca calcium	45 21 Sc scandium	48 22 Ti titanium	51 23 V vanadium	52 24 Cr chromium	55 25 Mn manganese	56 26 Fe iron	59 27 Co cobalt	59 28 Ni nickel	64 29 Cu copper	65 30 Zn zinc		70 31 Ga gallium	73 32 Ge germanium	75 33 As arsenic	79 34 Se selenium	80 35 Br bromine	84 36 Kr krypton
85 37 Rb rubidium	88 38 Sr strontium	89 39 Y yttrium	91 40 Zr zirconium	93 41 Nb niobium	96 42 Mo molybdenum	99 43 Tc technetium	101 44 Ru ruthenium	103 45 Rh rhodium	106 46 Pd palladium	108 47 Ag silver	112 48 Cd cadmium		115 49 In indium	119 50 Sn tin	122 51 Sb antimony	128 52 Te tellurium	127 53 I iodine	131 54 Xe xenon
133 55 Cs caesium	137 56 Ba barium	139 57 La lanthanum	178 72 Hf hafnium	181 73 Ta tantalum	184 74 W tungsten	186 75 Re rhenium	190 76 Os osmium	192 77 Ir iridium	195 78 Pt platinum	197 79 Au gold	201 80 Hg mercury		204 81 Tl thallium	207 82 Pb lead	209 83 Bi bismuth	210 84 Po polonium	210 85 At astatine	222 86 Rn radon
223 87 Fr francium	226 88 Ra radium	227 89 Ac actinium																

219

Acknowledgements

The publishers wish to thank the following for permission to reproduce photographs. Every effort has been made to trace copyright holders and to obtain their permission for the use of copyright materials. The publishers will gladly receive any information enabling them to rectify any error or omission at the first opportunity.

cover & p. 1 D. Roberts/Science Photo Library, p. 8t canismaior/Shutterstock, p. 8c Alex Mit/Shutterstock, p. 8b Science Photo Library/Alamy, p. 9t Tischenko Irina/Shutterstock, p. 9c koko-tewan/Shutterstock, p. 9b National Cancer Institute/Science Photo Library, p. 10 Picsfive/Shutterstock, p. 12 vladm/Shutterstock, p. 14 blickwinkel/Alamy, p. 15 Denis Pepin/Shutterstock, p. 16 Armin Hinterwirth/iStockphoto, p. 17 Biophoto Associates/Science Photo Library, p. 18 Graphic design/Shutterstock, p. 20 S.Pytel/Shutterstock, p. 22 Lisa F. Young/Shutterstock, p. 24t Graham Oliver/Alamy, p. 24bl Jubal Harshaw/Shutterstock, p. 24br Dr Jeremy Burgess/Science Photo Library, p. 26 Oyvind Martinsen/Alamy, p. 27 Artur Synenko/Shutterstock, p. 28 PHOTOTAKE Inc./Alamy, p. 29 Hank Morgan/Science Photo Library, p. 30t Dave Long/iStockphoto, p. 30b CNRI/Science Photo Library, p. 31 Sovereign, ISM/Science Photo Library, p. 32t Julian Gutt/Alfred Wegener Institute, p. 32b National Cancer Institute/Science Photo Library, p. 34 Richard Bizley/Science Photo Library, p. 35 Martyn F. Chillmaid/Science Photo Library, p. 36 shutswis/Shutterstock, p. 44t Sebastian Kaulitzki/Shutterstock, p. 44c Chad McDermott/Shutterstock, p. 44b Francois van Heerden/Shutterstock, p. 45t beerkoff/Shutterstock, p. 45c Nagel Photography/Shutterstock, p. 45b tehcheesiong/Shutterstock, p. 46 Rick & Nora Bowers/Alamy, p. 48 BSIP SA/Alamy, p. 50 photoseeker/Shutterstock, p. 51 Gail Johnson/Shutterstock, p. 52 Dmitry Lobanov/Shutterstock, p. 53 Jim West/Alamy, p. 56t Peter Treanor/Alamy, p. 56b Tree Frog/Alamy , p. 57t Angela Waite/Alamy, p. 57b Midland Aerial Pictures/Alamy, p. 58t Neil Bradfield/Shutterstock, p. 58b Nigel Hicks/Alamy, p. 59 Michael Willis/Alamy, p. 60t PJF/Shutterstock, p. 60b Horia Bogdan/Shutterstock, p. 62t Agripicture Images/Alamy, p. 62b Magdalena Bujak/Shutterstock, p. 63l Brian A Jackson/Shutterstock, p. 63r krugloff/Shutterstock, p. 63b Muellek Josef/Shutterstock, p. 64t Nathan Benn/Alamy, p. 64b Natural Visions/Alamy, p. 66 Flashon Studio/Shutterstock, p. 74t koya979/Shutterstock, p. 74u canismaior/Shutterstock, p. 74l PhotoAlto/Alamy, p. 74b Jan Will/iStockphoto, p. 75t fzd.it/Shutterstock, p. 75c Chepko Danil Vitalevich/Shutterstock, p. 75b Rob Byron/Shutterstock, p. 76 Alex Segre/Alamy, p. 77t Science Photo Library, p. 77b Phil Degginger/Alamy, p. 78 David Parker/Science Photo Library, p. 79 United States Department of Energy/Wikimedia Commons, p. 80t Martyn F. Chillmaid/Science Photo Library, p. 80c Charles D. Winters/Science Photo Library, p. 80b sciencephotos/Alamy, p. 82t Mark Evans/iStockphoto, p. 82b Niels Gerhardt/Shutterstock, p. 83t Andrew Lambert Photography/Science Photo Library, p. 83c Lyroky/Alamy, p. 83b Mathieu Viennet/Shutterstock, p. 84t Levent Konuk/Shutterstock, p. 84b Andrew Lambert Photography/Science Photo Library, p. 85 Andrew Lambert Photography/Science Photo Library, p. 86t sspopov/Shutterstock, p. 86b chris2766/Shutterstock, p. 87 Nomad_Soul/Shutterstock, p. 88t Jiri Hera/Shutterstock, p. 88b Losevsky Pavel/Shutterstock, p. 89 Mark Burnett/Science Photo Library, p. 90 PhotoSky 4t com/Shutterstock, p. 92 Charles D. Winters/Science Photo Library, p. 94 Evgeny Dubinchuk , p. 96 Pascal Goetgheluck/Science Photo Library, p. 97 Zoran Karapancev/Shutterstock, p. 98 Monkey Business Images/Shutterstock, p. 106t Sabine Kappel/Shutterstock, p. 106u AGphotographer/Shutterstock, p. 106l ggw1962/Shutterstock, p. 106b Master3D/Shutterstock, p. 107t PHOTOTAKE Inc./Alamy, p. 107c Adrian Sherratt/Alamy, p. 107b Charles D. Winters/Science Photo Library, p. 108t Monica Wilde 2011 www.angelflames.com, p. 108b x5 Andrew Lambert Photography/Science Photo Library, p. 109t x6 Andrew Lambert Photography/Science Photo Library, p. 109b Sherwood Scientific Ltd www.sherwood-scientific.com, p. 110t Alexis Rosenfeld/Science Photo Library, p. 110b Andrew Lambert Photography/Science Photo Library, p. 111t Charles D. Winters/Science Photo Library, p. 111b Alain Pol, ISM/Science Photo Library, p. 112 Radomir JIRSAK/Shutterstock, p. 114 Ivanova Inga/Shutterstock, p. 116t Givaga/Shutterstock, p. 116b Fancy/Alamy, p. 117t Martyn F. Chillmaid/Science Photo Library, p. 117b Martyn F. Chillmaid/Science Photo Library, p. 118t meunierd/Shutterstock, p. 118b US Department of Agriculture/Science Photo Library, p. 120 Editorial Image, LLC/Alamy, p. 122 robertosch/Shutterstock, p. 123 Charles D. Winters/Science Photo Library, p. 124 Gertan/Shutterstock, p. 126 komar_off/Shutterstock, p. 134t Bryan Solomon/Shutterstock, p. 134c Jan van der Hoeven/Shutterstock, p. 134b Faraways/Shutterstock, p. 135t dusan964/Shutterstock, p. 135c Ken Hurst/Shutterstock, p. 135b jokerpro/Shutterstock, p. 136 Du Cane Medical Imaging Ltd/Science Photo Library, p. 137 Monkey Business Images/Shutterstock, p. 138t Hugh Lansdown/Shutterstock, p. 138b Larry Mulvehill/Science Photo Library, p. 139 jovannig/Shutterstock, p. 140 Jo Ann Snover/Shutterstock, p. 142 Quayside/Shutterstock, p. 144 michaeljung/Shutterstock, p. 146 StockLite/Shutterstock, p. 148t Andrew Buckin/Shutterstock, p. 148c Alan Crawford/iStockphoto, p. 148b Kosmonaut/Shutterstock, p. 152 gresei/Shutterstock, p. 154t Andy Green/iStockphoto, p. 154b saiko3p/Shutterstock, p. 156t Alexis Rosenfeld/Science Photo Library, p. 156b Ljupco Smokovski/Shutterstock, p. 157 Seb Rogers/Alamy, p. 158 Imagestate Media Partners Limited - Impact Photos/Alamy, p. 159 muzsy/Shutterstock, p. 160t ImageState/Alamy, p. 160b Petr84/Shutterstock, p. 162 Lagui/Shutterstock, p. 170t Pi-Lens/Shutterstock, p. 170c mamahoohooba/Shutterstock, p. 170b Albert Lozano/Shutterstock, p. 171t Anton Kozlovsky/Shutterstock, p. 171u PeterPhoto123/Shutterstock, p. 171l Ivaschenko Roman/Shutterstock, p. 171b HLPhoto/Shutterstock, p. 172 Brandon Blinkenberg/Shutterstock, p. 174 estike/Shutterstock, p. 176 Eye-Stock/Alamy, p. 178t Nick Hawkes, Ecoscene/Corbis, p. 178b Andrew Lambert Photography/Science Photo Library, p. 179 Masterpiece/Shutterstock, p. 192t Andrew Lambert Photography/Science Photo Library, p. 192c Pedro Salaverría/Shutterstock, p. 192b Shawn Hempel/Shutterstock, p. 198 Martyn F. Chillmaid/Science Photo Library.